PREFACE

After a walk on a cold autumn afternoon, I was sitting in the house watching the fire along with my dog Jaeger, a Deutsch Drahthaar. From the wood pile next to the fireplace, a deer mouse emerged and scampered over the floor and headed to another room. Comically, at least to me, when the mouse hit the wood floor his feet went every which way while it struggled to get traction. Jaeger and I were too tired to give chase. Instead it brought back the sounds and sights of things we'd seen that afternoon. A pileated woodpecker making its long, oblong holes just off the ground in a dead tree. A late garter snake looking for an overwintering spot. The ever present red squirrels, chattering their disapproval at our presence. And high in a tree, a porcupine, oblivious in its coat of spines. Who knows what was out there that we didn't see. Cougar?

There is much to know about the north woods, and I confess to being a novice on most topics, even though I earned a Ph.D. in zoology. For example, I have a neurological block about plants. If you show me a plant and tell me its name, five minutes later I will swear that I've never seen it before. Maybe it's my partial red-green color blindness that makes plants, which I'm told are mostly green, something that doesn't leap out of the background for me. True, I know the names of a few trees, and I'm decent at recognizing poison ivy and parsnip in bloom, but mostly I just don't know plant names. I consider it a scientific character flaw but am now content in the realization that I'm too old to do anything about it.

The same is true about most things that live beneath the surface of the water. Just looking out over the lake, it looks mostly devoid of life, save for the occasional gull or loon. It is easy to forget that there is an abundance of life beneath the surface that we don't see unless we really look. Sure, I know the fish I try and catch, and the invasive rusty crawfish we trap off our dock (and cook up), but the many smaller fishes and the even smaller microorganisms that form the basis of the food chain are pretty much a mystery. I guess it reinforces out of sight, out of mind.

Insects are, however, fascinating. I once had a summer job at the University of Minnesota's Itasca Field Biology Station, but I had to be enrolled in a summer class. Arriving to the station at the last minute, there was but a single open spot in any class, and that was entomology. Argh, I was pretty sure I hated insects but bugs it was. It took only a day of capturing insects in a net, killing them humanely, and looking at them under a microscope to reveal their incredibly complex anatomy. But I had already fallen in love with birds.

To provide perspective to my outdoor life, I once took our youngest son to a marine dealer, and we fell in love with a nice boat. I said, well we can buy this boat or I can take your mom on a vacation to Provence, in southern France. He reflected a moment and said in all seriousness, "Mom loves to fish." That was 3 boats ago.

I put these essays together thinking about cold winter days, rainy days, days with the wind and rain keeping you off the lake and out of the woods, in other words, a day inside the cabin. I hope they bring you a smile and some new ideas. Each of the chapters begins with a section entitled Cabin Talk -- an overview of the topic to enable a reader to decide if they want to spend 3 minutes reading! I hope some of the things I have written will make you say, "That can't be true" and that you'll look it up to see whether I was right! I do that all the time!!

CONTENTS

V. DOGS, BUGS, AND STUFF   303

GET TO KNOW THE DEER FAMILY

1. Are There Too Many Deer?

Cabin Talk.—You're sitting at the table in the late afternoon and notice movement in the woods near the edge of your lawn. It's a deer starting its nightly ritual. You reflect on a recent discussion with a friend. The topic was, are there too many or too few deer? Car insurance companies think there are too many, as do some environmental groups. In my experience, it's impossible to find someone who says the number of deer are just about right.

An often-heard concern is that there are too many deer and they do tremendous ecological damage, causing some to call for reduction in deer numbers. On the other hand, I don't think I've ever heard a hunter say "Yup, there sure are too many deer." These two conflicting views on the size of the deer herd put wildlife managers squarely in the crosshairs from every direction. How large should the deer population be in a given area?

USDA Forest Service biologist Dr. Brice Hanberry published some thought-provoking ideas arguing that there are in fact not too many deer. I am always reminded that 21,000 years ago much of northern North America was covered by a mile thick glacier. What I have not concentrated on, and Dr. Hanberry reminded me of, are the many large animals that were present on our landscape until about 12,000 years ago, when most of the "megafauna" went extinct. Let's take a step back in time.

If you went on a field trip 12,000 years ago across much of eastern North America you would have encountered: an armored armadillo-like creature called a pampathere (500 lbs), an actual armadillo with a tail spike called a glyptodont (1,800 lbs), an elephant-like beast called a gomphothere (10,000 lbs), a couple of horses (one up to 3,000 lbs), a tapir (700 lbs), several species of ground sloth (up to 6,000 lbs), capybara (200 lbs), mastodon (12,000 lbs), mammoth (13,000 lbs), American zebra (700 lbs), several peccaries, three species of llama (up to 800 lb), giant bison (4,000 lbs), an extinct pronghorn, giant beaver (170 lbs), an armadillo (twice the size of the living species), bison, among others. White-tailed deer and mule deer were also present along with at least one other extinct cervid species.

By the way, with all that meat on the hoof there was a scary set of predators, including American cheetah, American lion, dire wolf, jaguar, saber-toothed cat, scimitar-toothed cats, and short-faced bear, and a host of scavengers, including several extinct vultures. So, for the sake of argument, if you had been alive then, you'd not want to have been the slowest one in your clan.

Dr. Hanberry emphasized three points. One, these are all herbivores, meaning they are grazers or browsers, and second, many were big, really big. The big ones are called "megaherbivores." When you add up the sizes of these animals, it is clear that being a plant would have been dangerous, as so many huge things wanted to eat you and all of your friends. Now, we know that they did not all live in the exact same place (some were only in the southern US) or eat the exact same diet. The third point was that bison, deer (and pronghorn in the west) were the only megaherbivores to have survived the extinction event of 12,000 years ago.

Why did the megaherbivores and their friends perish from North America? Scientists have long debated the causes; one explanation involves climate change to which animals could not adapt fast enough. The other is the so-called Overkill Hypothesis, in which the growing human population drove all the megaherbivores to extinction by hunting. As with many concepts in biology in which people argue one or the other, I'm guessing it was some of both.

The megaherbivore in the proverbial room is what ate all that vegetation once the big eaters were gone? Obviously, the surviving deer couldn't do the job of tons of megaherbivores. Lots of animals, insects in particular, are herbivores, and perhaps the extinction of the megaherbivores opened major new ecological frontiers for them. Although if you're a dung beetle, your glory days have passed. We know that loss of species from an ecosystem can alter the number and abundance of plant species because some of the now-missing species were seed dispersers whereas some created ecological opportunities for other species. I wondered what would have happened to the vegetation of North America 12,000 years ago when all those huge plant-eating species went extinct? Was it a time of Plants Gone Wild?

Dr. Hanberry believes that after the megaherbivore extinctions open woodland habitats were maintained by frequent and extensive fires. That is, fires probably weren't extensive during the reign of the megaherbivores because they ate most of the fuel. But upon their demise, fires replaced the ecological services of megaherbivores. However, she also notes that the oak woodlands included some fire-resistant species, meaning that fires were not unheard of, just not as frequent as they became post-megaherbivores.

Dense closed forests were encountered by early European settlers, especially in the upper Midwest and New England, but Dr. Hanberry puts it this way: "Many historical ecosystems were open forests (savannas and woodlands) with low tree densities, which allowed greater diversity and abundance of the herb layer." This sort of environment is good for deer and may account for the high numbers of deer pre-settlement. Then with fewer fires the forest ecosystem changed from savannas and woodlands to more closed forests in some areas. Incidentally, the upper Midwest and New England probably did not undergo drastic changes other than those caused by logging. Deer numbers crashed during the 1800s to very low numbers in the early 1900s as a result of over harvest although populations have rebounded to previously high numbers (around 1600).

Dr. Hanberry would prefer to see the landscape return to more open forests and having more deer would help (keep trees from getting very big). We like closed forests for the wood they provide, although high deer numbers often result in harm to forest understory, which is why some favor fewer deer. From the standpoint of restoring the historical ecological landscape, Dr. Hanberry, and many hunters, would like to see more deer. Balancing these concerns makes the deer manager's job about like herding angry wasps. There is, then, little consensus on whether we have too many or too few deer.

2. Are There White-Tailed Deer On Other Planets?

Cabin Talk.—Another day, another early morning. You see deer getting in a last mouthful before bedding down for the day. In his book "Astrophysics for people in a hurry" Neil deGrasse Tyson wrote that "In the beginning, nearly fourteen billion years ago, all the space and all the matter and all the energy of the known universe was contained in a volume less than one trillionth the size of the period that ends this sentence." As the last of the stars fade into the light of day you wonder about the galaxy.

Because no one can hear your thoughts, you chuckle to yourself and ask whether deer exist on other planets. Sure, it's a dumb fantasy, but some of our most intriguing insights start with a far-fetched fantasy.

I sometimes ponder my miniscule and insignificant role in the universe, and how vastly little we know about our place in space and time. Tyson's quote about the universe takes a while to sink in. But even if the scope of the universe is outside the realm that most of us comprehend, it needn't stop us, and in fact should encourage us, to think "outside" the box, maybe way outside.

How's this for outside the box: Are there deer on other planets? That's a stupid question you say, and I would agree. But there have been many scientific and popular articles on the existence of extraterrestrial life, so why not ask whether there are deer or deer-like creatures on other planets? I admit that the difference between the probability of deer on other planets, and zero, is about zero. Scientists, however, never say never.

When asked about the existence or reality of UFOs, Neil deGrasse Tyson said something like, in UFO the first letter represents the word "unidentified." So if it's unidentified, we have no idea whether it came from central Russia, Area 51, or 70 trillion planets away. Tyson also remarked that he has seen no evidence of visitors from extraterrestrial places. And before someone wonders if he's involved in a massive coverup conspiracy, remember that he would be even more famous if he made public any scientific evidence of extraterrestrial life. Furthermore, ask yourself whether there exists, or has ever existed, a news reporter who would keep that information secret. If I told a reporter that this is "off the record" but I showed them unambiguous proof of life on other planets, it's my bet that the reporter would rather go to jail than keep that secret.

A simpler question than deer on other planets, is whether there is "life" elsewhere in the universe. Of course this begs the question of what is "life" ? NASA defined life as "A self-sustaining chemical system capable of Darwinian evolution." One sometimes hears that evolution must be false because it cannot explain the origin of life. However, evolution refers to changes in living beings via natural selection or other evolutionary processes after life arose, and not the origin of life itself. Even NASA's broad generalization assumes that "life" is something we'd recognize based on our earthly experiences. We often hear that our search is for "intelligent" life, but maybe a more defensible search might be for "alien life," with a broad definition of life. Is life solely on our planet, and if so, how could it arise?

Many scientists have explored ways in which life might have originated. One of the older ideas that has garnered additional support is called the "panspermia" hypothesis. It suggests that the precursors of life as we know it hitch rides on meteorites, of which over 15,000 hit earth every year. Probably every planet is hit with cosmic debris. Hence, there are likely billions of "early life experiments" in progress throughout the universe. Sound far-fetched?

Our genetic alphabet encoded in DNA and RNA, consists of just a few "bases" including adenine, guanine, cytosine, thymine, and uracil (RNA). When strung end to end these form the instructions or blueprints for how individual organisms develop. Each of these "nucleobases" have now all been found in meteorites that have crashed to earth. In short, some of the basic ingredients for life as we know it might have come from extraterrestrial sources and if these meteorite-bound essential life ingredients encountered the right conditions on early earth, it might have started the proverbial Primordial Soup boiling. Clearly, if these nucleobases have arrived from extraterrestrial sources, life even as we know it might not be so far-fetched after all. But is there any evidence that we are not alone in the universe?

We often hear that we have spent a lot of time monitoring space for signs of alien life and found nothing. Does that answer the question in the negative – there is no life elsewhere? Any scientifically minded person would ask a simple question. How much of the universe have we "sampled"? Dr. Jill Tarter, an astronomer with the organization SETI ("Search for Extraterrestrial Intelligence") put it this way. Set the volume of the universe equal to all of the water in earth's oceans, and ask the question are there any fish in the oceans. The amount of exploration and monitoring of the universe for extraterrestrial life that we have done to date is equivalent to taking one 12 oz glass of water from somewhere in earth's oceans (352,000,000,000,000,000,000 gallons) and, seeing the glass full of nothing but water, conclude that there are no fish. We have a sampling problem at the universe level.

As I noted above, the probability of an earth-based life system replete with ungulates like white-tailed deer is, I would guess, about as close to zero as you could get. However, consider the immensity of the universe. One estimate of the number of planets in the universe is 100,000,000,000,000,000,000,000 (that's 1 followed by 23 zeros). If the probability of a deer-like creature evolving on another planet is 1 in a 100 quadrillion (100,000,000,000,000,000), we should eventually find a million planets with deer or deer-like ungulates. Of course we might find a planet with living things that ate other living things (however living is defined), and we might consider that a win. Although I admit to rolling my own eyes at the thought of ungulates in space, I'd love to know what we find in the next 100 years of monitoring the universe. But then again, I guess that's everyone's lament.

3. Deer Birth Control?

Cabin Talk.—If there was a viable solution to a problem that mattered to you, but you didn't like the solution, would you support it? What if that solution required you to abandon your cherished activities? In particular, what if it were decided there were too many deer but hunters would not be allowed to reduce the herd? Are there workable, viable ways to reduce deer numbers without hunting?

There have been many attempts by anti-hunter's to stop hunting, even bow hunting, as a means of reducing the size of a deer herd, one that (not) everyone agrees is too big. It's easy to find areas where an overabundant deer population has caused ecological damage, such as removing native vegetation and paving the way for invasives that are unwelcome to us and the deer themselves. Most wildlife professionals recognize that hunting deer provides an important ecological service, while realizing that different people can each have their own opinions about whether they approve of hunting. But to oppose hunting as a means of population control, another equally viable ecological and economic solution needs to be presented.

The borough of Staten Island has a problem with too many deer. The usual problems have arisen, like deer-car collisions (103 in 2018) especially during the rut, and over 90 new cases of Lyme disease, an increase of over

250%. Not to mention residents complaining about deer eating flower beds and gardens, and even worse, eating taxpayer-funded and city-maintained greenery around parks.

With hunting forbidden, the local deer population grew from about 24 in 2008 to over 2000. What to do? Other cities have tried birth-control in the form of contraception drugs, which as I reviewed earlier, didn't work. You have to make sure the drugs do not get into other animals, does have to be regularly retreated, and you can't prevent deer from neighboring areas from wandering into the treatment area.

Staten Island decided to try a different procedure, vasectomies. They are probably the only city to try this approach. A captured buck is anesthetized, and a vet does the snip job in about 15 minutes. After that, Randy becomes Randie, and the population should start to decline.

Most would argue that you can't catch enough bucks to be effective, and it only takes a few non-sterilized bucks to mate with a lot of does. However, officials in Staten Island claimed they have caught quite a few, up to 92%. The goal they set is 98% of the bucks on the island. They are searching for the missing 250 bucks they estimate to be on the loose; I send those bucks my best wishes for escape from the vets. And guys, avoid the bait sites, the dudes with dart guns await you there.

So, what does this program end up costing? The city hired a non-profit group called "White Buffalo" to do the snipping. The city has spent about $4.1 million over three years. They performed vasectomies on 1,577 deer, and the population decreased by about 316 deer, or 15%. One way of looking at it, the city spent $12,975 per head to reduce the population by 316 deer (15%). City officials argue that the goal of a reduced herd will be seen in the coming years when those sterilized bucks won't be breeding. In that case, the cost is still $2600 per deer sterilized. And, White Buffalo negotiated an extension of their contract at a cost of another $2.5 million dollars. Hunters are saying, rightly in my opinion, "Heck, I'd do it for free."

Although we hear about government spending problems, cost overruns and budget issues even affect the deer vasectomy business. The city said that the original budget was not to exceed $3.3 million (as if that was supposed to be cheap). However, White Buffalo found more deer than expected, and they kept asking for more money, eventually reaching the $4.1 million figure. What does the money go for?

A Freedom of Information request showed that only 7.6% of the $4.1 million ($311,600) went to supplies and bait, the rest was used to pay "senior scientists, wildlife biologists and technicians, and veterinarians charged with capturing and sterilizing the deer." According to a local newspaper who obtained the information, "Many of those scientists earned six-figure salaries often for less than half a year of work, and in some cases, earned more than the median household income of New York City residents for just a month of work."

Some City officials declared success, at least in the long term. However, Staten Island Borough President James Oddo remarked "The beneficial impact of their method is still years away," "Meanwhile, crashes, increased incidence of [Lyme] disease, and the extinction of plant species and various ecosystems roll on." Oddo has pushed the city to bring the deer numbers down with a controlled hunt. "The hard reality of what we must do remains right there in plain view in front of us," he said. "It's time we acknowledged it." However, this requires cooperation and approval from both state and federal agencies, not an easy task.

What happened when Oddo brought groups together was that the New York Police Department and the city council were against it, likely for the usual political reasons (they want to get re-elected). Furthermore, hunting is illegal in all the boroughs, and any cull hunt would be delayed by countless lawsuits brought by activists. The Parks department put it like this: "A cull was considered unrealistic because Staten Island is still an urban environment and even a controlled slaughter would require large swaths of borough green space to be cordoned off by the NYPD." How many anti-hunters do you think were on the parks board to term a cull hunt a "slaughter"?

Maybe deer hunters shouldn't sell out for cheap — how about they pay hunters $5000 a deer? I'm thinking there'd be a big line at the sign-up place.

Lastly, Oddo is checking out a new technique called "deer avoidance technology." Here, detection devices reflect motor vehicle headlights and give off a high-pitched noise to scare deer from roads when cars approach. Ironically, at least to me, that keeps more deer alive. Of course, that's good for residents and car insurance companies, body shops would argue the opposite. The technology is currently unproven, I'm dubious but could be convinced.

Perhaps the best view of the vasectomy program was written by Bernd Blossey, an ecologist at Cornell University: "It's difficult for me to come up with all the reasons why this is a really stupid plan." A cull hunt would do the job but the obstacles might be insurmountable.

4. What Do Deer And Bald Eagles Have In Common?

Cabin Talk.—Although much of news about human effects on the environment is negative, there have been some success stories, stories that hopefully encourage us not to give up on trying to reverse negative things we've done in the past. Sometimes our view from the cabin is that there's not much any individual can do, which is why an environmentally educated population is a prerequisite to any future environmental wins.

The U.S. has seen conservation successes and failures. Whooping crane, California condor, American bison, peregrine falcon, and bald eagle are successes, so far. Not so lucky were the passenger pigeon, Carolina parakeet, ivory-billed woodpecker and Labrador duck.

North American deer are another example of a success story because of human intervention. Once, there were large numbers of white-tailed deer along the borders of the eastern forests, mule deer and elk to the west. The deer we now call Coues deer was in the Sonoran Desert and the Key deer in Florida. White-tails were not common in the dense forests of the east because there wasn't much understory vegetation to provide food. Native Americans regularly used prescribed burns to open land for planting crops and to prevent sneak attacks on their villages. When the plot ran out of nutrients, the tribe moved on and the fallow fields became good for deer.

When colonists started clearing land for crops, they did it on a larger scale and kept it from reverting to forests. This increased deer populations. This increase, however, did not go unnoticed by colonists and Native Americans alike, and hunting of deer increased dramatically. According to J. B. Trefethen, it was easy to take six deer a day. People made a good living selling venison and deerskins. The overharvest of deer became so bad that by 1646, Portsmouth, Rhode Island, closed the open season on deer from May 1 to November 1, and imposed a fine of 5 pounds ($1219.38 in today's dollars) for killing one. Other cities followed suit. The commonwealth of Virginia noted: "many idle people making a practice, in severe frozen weather, and deep snows, to destroy deer, in great numbers, with dogs, so that the whole breed is likely to be destroyed in the inhabited parts of the colony." As a result, no one could kill any deer from 1772 to 1 Aug. 1776, and if you did and were caught, you would be flogged. I digress, but I wonder what the job description was like and how much a flogger earned? But let's move on.

Few deer regulations were enforced, and people at the frontier continued to take deer for food. Then came the market hunters, and because of new travel opportunities made possible by roads and railroads to eastern markets they had a severe negative impact on the deer herd. There was the Great Hinckley Hunt of 1818 – not the Hinckley in MN known for the big fire of 1894. Rural residents of Medina County Ohio were losing livestock to wolves and bears and organized a Christmas hunt in which 600 men and "large boys" took to the woods. The group lined up in a square and at sunrise, the commander shouted "all ready" to his right and the hunters passed on the cry, which traveled around the square in 40 seconds! Bugles sounded, dogs barked, and the procession began with the lines moving towards the center of the square, the crew commenced shooting and clubbing anything they could. Once the four sides got too close to each other, they climbed trees and shot downward; amazingly, only 2 hunters were wounded. The toll was 17 wolves, 21 bears,

300 deer and lots of foxes, raccoons and turkeys. A grand holiday feast ensued. Those were the days.

The point was that no one cared that they wiped out the local deer population. By 1859, deer were gone from Iowa. The market hunters moved west and took on bison, pronghorn, elk, mule deer and bighorn sheep. Nothing was safe.

The situation changed because of logging of the northern coniferous forests, which created more deer habitat. Where there had been few deer in northern Minnesota, Michigan, Wisconsin, New Hampshire and Maine, by 1870 they were common. The (over) harvest was back on. In December 1872, it is reported that six tons of dressed venison were shipped from Litchfield, Minnesota, to markets in Boston. Granted, at 50 lbs per deer, that's only about 250 deer, but it's for one month. In 1880, 100,000 deer were shipped from Michigan to the east coast. By 1890, the North American deer population was in bad shape. The following states had almost no deer: Rhode Island, Connecticut, Maryland, West Virginia, New Jersey, Ohio, Kentucky, Tennessee, Indiana, Illinois, Iowa, Kansas, Missouri, and Nebraska. Other states had deer, but not in great numbers and these states began to notice what was happening elsewhere – in 1873, Maine was the first state to have a bag limit, three per hunter per season. Other states followed suite.

There were a few pockets of deer in wilder areas, and in 1890 the ornithologist TS Palmer estimated 300,000 total deer in the US and Canada. But at this time, land use changes had once again started to favor deer and at least in northern Maine and New Hampshire, where the wolf was now facing extinction, deer populations increased, aided by bag limits and seasons.

Finally waking up, states formed agencies whose job it was to protect wildlife. By enforcing bag limits the market hunter was out of business. From surviving pockets of the remaining deer herd, the population expanded in all directions. By 1908, the deer herd east of the Mississippi was thought to be 500,000. Another factor contributed to the recovery of deer. Captive deer were purchased and released into the wild. Some states purchased deer and released them in state forests. For example, the Pennsylvania game commission started stocking deer in 1900, and by 1907, hunters harvested 200 bucks. Clearly, given the chance, deer populations can rebound rapidly.

We want more deer, say some. We should be careful what we wish for. By 1925, deer were everywhere in Pennsylvania, and large herds of 40 or more were common. However, shortly after, biologists noted stunted antler growth, and some hunters reported seeing 100 bucks but only spikes, which were illegal. In winter 1926 deer died by the thousands from starvation due to overpopulation. A biologist counted more than 1000 dead deer in four townships of a county, a hefty toll.

So what was the answer? If you only harvest bucks bigger than a spike, which can breed, the population continues to increase and you have no choice but to open a season on does. This is probably the origin of antlerless deer harvest, and between 1931 and 1941, hunters harvested more than 735,000 deer in the woods of Pennsylvania.

The return of deer didn't happen as quickly in other places as it did in Pennsylvania. The Great Depression had two effects, one, people left their farms, which aided deer, and two, hungry people shot deer irrespective of laws. A landmark piece of Federal legislation was passed in 1937, the Pittman-Robertson Act. This resulted in an 11 percent excise tax on hunting equipment being returned to states to be used only for its wildlife agencies. Much of this revenue went to help deer populations and the herd exploded nearly everywhere. Survival increased and transplants aided population growth. By 1965, Kansas opened a deer season, which made every state east of the Rockies a big-game state. By 1968, hunters harvested 1,500,000 deer, more than twice the population a half-century before. By 2000, hunters killed 7,500,000 deer, out of an estimated population of 40,000,000.

From 300,000 to 40,000,000 is quite a conservation success story. The story shows a remarkable history of rapid up and down trends, near extinction, and marked recovery. Conservation efforts played a huge role in both the recovery of deer populations and the bald eagle. I like seeing eagles, and I'm very glad they're still around. I like deer too, just check my freezer. (Much of the material in this article came from a 1970 article by James B. Trefethen.)

5. Deer With Hairy Eyeballs, The Case Of Damn Unlucky

Cabin Talk.—Freaks of nature are the stuff of traveling shows of a bygone era. About all we could say about a seven-legged lamb, a giraffe with a 90 degree bend in its neck, a cow with two legs growing from its shoulders, a one-eyed kitten, or two conjoined crocodiles, is oops, mother nature screwed up. Some are relatively easy to understand, like the three-eyed fish found in a nuclear power plant reservoir. Today, biologists who study embryological development, called unsurprisingly developmental biologists, have learned how these aberrations come about.

They might not have a preventative solution, but they understand what goes wrong to yield mother nature's oops.

Early one morning I saw a headline in the National Deer Association online newsletter that read "Freak buck had corneal dermoids." In my pre-caffeinated daze, I thought to myself, why would a buck have cornmeal in his, wait, what are dermoids? The article by Lindsay Thomas Jr. was, in a word, fascinating.

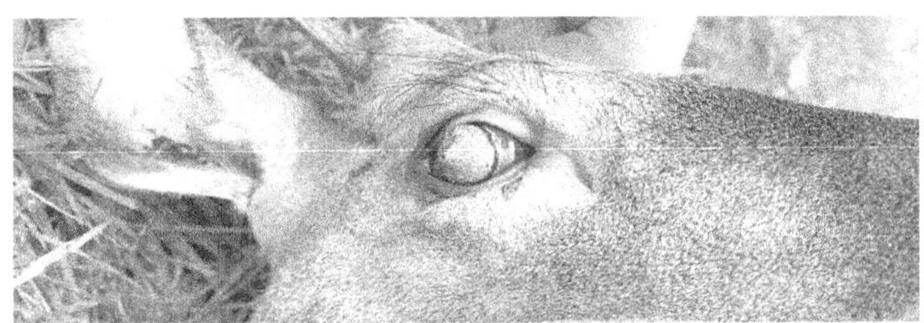
Unlucky deer with corneal dermoids and Epizootic Hemorrhagic Disease Image - Tennessee Wildlife Resources Agency

I had just finished lecturing about the human genome, and how cool it is that in all, or most, of our cells we have the DNA blueprint to make anything we want, anywhere on our bodies. But there is genetic regulation at the cellular level. For instance, the genes for making fingernails are only turned on in the cells at the tips of our fingers (and toes), muscle cells in the right places, and genes for our complex eyes in the sockets in our skulls. This process is a matter of gene regulation, so that we don't have fingernails on the tops of our heads, eyeballs on our knees, etc. Hey, to digress on an off-the-wall idea, this might be useful in baseball umpires.

The process, however, sometimes goes awry. In the fruit fly, some mutations occur that result in the antenna being turned into legs – cool, but not if you're that fly. Other flies can have two sets of wings. Sometimes the developmental process gets really screwed up, and we've seen a cow with two heads, turtles with heads at each end, and people with tails. Yes, some human infants have tails (fewer than 50 known cases), which are usually surgically removed. The term for expression of genes and growth of the wrong structure in an abnormal place is an atavism.

The freak buck mentioned above, who I named "Damn Unlucky," (DU, with apologies to Ducks Unlimited), was shot as a one and a half old in eastern Tennessee in late August of 2020, as it was wandering around in a daze, obstructing traffic and attracting attention. No wonder DU was wondering around - he had hair growing all over his eyeballs. Remember that vertebrate eyes have an iris (the colored part), a pupil (the dark, open center), and a lens, all of which are covered by the cornea, that thing if you scratch hurts like the devil. The tissue in DU's cornea was the sight of the accidental expression of skin genes, which did their due diligence and made hair, like they're supposed to, just not there.

Contributing to DU's condition was EHD, epizootic hemorrhagic disease. Unlike CWD, in fact very much unlike, EHD is caused by a virus that is spread by "no-see-ums." Populations are often hit very hard by EHD, but the viral infection in the population goes away after time. CWD, caused by a pool of ever-increasing abnormal protein called an infectious prion, a nonliving entity, builds up in the soil like an environmental toxin and keeps reinfecting deer irrespective of how many deer are in the area. A deer's genotype can impede the onset of the disease but it is considered 100% fatal at present. EHD and CWD are very much unalike. Still, DU had EHD and showed the symptoms of disorientation, much like deer with CWD. Whether DUs eyeball issue was related to EHD is unknown. It certainly couldn't have helped his vision. Lots of deer have had EHD with no reports of accompanying corneal dermoids (see below).

The term for this eyeball atavism is a corneal dermoid. The term dermoid is derived from dermis (skin), and is a piece of functional skin growing where it would not be normally, in this case the cornea of DU. The genes can produce glands and other features in addition to skin. Many other animals have had this affliction, including dwarf rabbits, hairless guinea pigs (now that's ironic), cows, and at least one human.

Mr. Thomas reports that a pathologist suggested that DU had the affliction from birth and likely survived for as long as it did because it was in a relatively safe urban area and attended at least early on by its mother. It's not clear whether the dermoid increased in coverage and severity over time, or whether DU was effectively blind from birth

Damn Unlucky is apparently one of only two known white-tails to have had this affliction. The other known white-tail with corneal dermoids was a 1.5 yr old doe harvested in St. Mary parish Louisiana in 2007 and was described in an article by LaDouceur and colleagues in the Journal of Wildlife Diseases. Unlike DU, this doe had 3 relatively small spots on the left eye only. Upon dissection, it was found that these three masses on the deer's cornea were comprised of hair follicles, sebaceous (oil producing) and apocrine (sweat) glands, and well-differentiated cartilage.

Most surprisingly to me, the authors reported that they found a cystic tarsal gland in the eyelid! Talk about misplaced genes! A tarsal (leg) gland showing up in an eyelid. That's a major oops.

These corneal dermoids are stark reminders of how well coordinated the vertebrate developmental systems are – going from a single fertilized cell to millions of cells all genetically programmed to develop in the correct places to produce a complex body like a white-tailed deer, with only two known instances, out of millions of harvested deer, of hairy eyeballs. So, on average, the genetic program directs development in the correct way.

6. Tusks And Antlers: Now You See Them, Now You Don't

Cabin Talk.—Many non-hunters seem to assume that big game hunters are all after trophies. It usually escapes their attention that the meat from every large buck they see mounted on a wall was eaten, albeit I'd rather have a doe because I think that large bucks are kinda tough and not so good tasting. Not that I have a large sample size (see blessed bucks). I've had exactly two bucks mounted, one of which I gave away to our movers. But no more for me, because I realize that a couple of weeks after I'm gone, my remaining buck will be in a garage sale for a couple of bucks. So why shoot bucks unless they're worthy of wall hanging? By protecting more does, it benefits the population. Still, the sight of large-antlered deer leads to what is termed buck fever.

Imagine a buck deer without antlers, a ram without horns, roosters without bright colors and long tails, or toms without beards. Those animals would certainly lend a different perspective to some hunters. To me, not so much, as most of those "exaggerations" are not edible. Now, if walleyes never exceed 12 inches, that could be a problem.

What about an elephant without tusks? I'm personally not interested in hunting elephant hunt, for a number of reasons, but it seems to me that elephants without tusks would be like shooting a bloated cow, hardly seems fun and certainly there's no place on the wall. Maybe sometimes our inner Tarzan peaks out.

Why the speculation? Hunting can change the way an animal looks. We know that because of (trophy) hunting the sizes of horns on breeding rams has dropped by 20% over 20 years. Size of some fish species has also decreased because of fishing pressure. These changes result in breeders being younger and less mature. Now of course rams with huge trophy sized horns were once less impres-

A tuskless female elephant in the Gorongosa National Park in Mozambique with two calves (from ElephantVoices.org).

sive and probably passed over by hunters, and one cannot predict a ram's future horn size. Given no choice hunters will shoot the largest one available, even if it doesn't pass muster by older standards, causing a shift to overall smaller horns in the breeding cohort.

Back to elephants. Both male and female elephants have tusks made of ivory. They are basically the same tooth as your incisors, although your incisors are not ivory. Typically, the older the male, the bigger the tusks. The largest tusks recorded were 226 and 235 lbs, respectively. A different elephant had tusks that were 11 ½ feet long!

Elephants have been poached for a long time, owing to the value of ivory. Ivory can fetch $1500 per pound on the black market, so a biggish male elephant with two 125 lb tusks is worth a lot to poachers. Tusks in females, while smaller, are also valuable.

In Mozambique, poaching of elephants during civil war strife (1977 to 1992) has led to the appearance of a fairly high percentage of females lacking tusks (see photo). Clearly, having tusks is like having a target on your back (kinda like wearing a shirt that says DNR). An interesting scientific paper was published in the prestigious journal Science by Shane Campbell-Staton (Princeton U) and colleagues entitled "Ivory poaching and the rapid evolution of tusklessness in African elephants."

They found that elephants poached for ivory during a long civil war resulted in a decrease in elephants (90% of the population was lost) and an increase in the proportion of females without tusks, which makes them safer from poachers. In fact, the proportion of tuskless females increased to 33% over about 30 years. Pretty clever for an elephant. But how exactly does an elephant go about not having tusks, and why do males not show this advantageous condition? Of course it's a mixed condition for males, as tusks help with being a dominant male, and provide a target for poachers.

The answer appears to be with the genetics of the sex chromosomes. Like humans, elephant females are XX and males are XY. It turns out that there are a couple of candidate genes on the X chromosome. Females with a particular mutation at one or both of two genes (called AMELX and MEP1a) will lack tusks. Males with that same mutation on their sole X chromosome will not be born. So, there are no tuskless males; bummer for bulls.

Sounds good for at least female elephants, but all is not well. It turns out that tuskless females have different diets than the genetically typical elephants, which influences the local ecology. In particular, tuskless females eat more grass and fewer woody plants, which are normally extracted from the ground with their tusks; being tuskless leads to a different diet. Elephants are considered "keystone" species, meaning they have a major impact on the vegetation in an area. Also, with the frequency of these genetic mutations occurring on the X chromosome, which are fatal to males, the overall population will decline owing to fewer males.

As an evolutionary biologist, I object to saying that tusklessness "evolved." Tuskless females were present (about 18.5% of females) before the intense culling, and poaching changed the frequency of tusklessness in female elephants, but it has not caused a new evolutionary innovation (e.g., like a third tusk). If a tuskless female has female offspring, whatever gene or genes that caused loss of tusks will increase in the population. However, the tide can turn. In the absence of intense harvest, I predict that the frequency of females with tusks will increase because tusks are useful in digging for water, stripping bark for food and physical contests with other elephants. That is, in the absence of poaching, tusked females will have the advantage over tuskless females. Thus, natural (or unnatural) selection has caused this shift in frequency.

What drives the price of ivory so high? It was banned in 1989 by the international community, and the ivory-carving industry in China was crippled. Only pre-ban ivory could be carved into trinkets and figurines, and with high demand, prices soared. In 1999 the carving industry rebounded, resulting in a lucrative market for poached ivory, as there was no way to tell when the ivory was obtained.

Lest we cast stones, or viruses, towards China, the US is the second-biggest ivory market. It is now illegal to sell ivory in the US unless you can show that it was obtained lawfully before 1990 and you sell it only within your state. Some states are stricter; in California it is illegal to sell any ivory within the state.

For me, killing an elephant to make ivory trinkets is pretty galling. In my opinion it reveals a lack of respect for an elephant for it to meet its fate in this way. Apparently in some parts of the world people think that tusks are like fingernails and will regrow if cut off. They are not - the lower third of the tusk is embedded in the skull and the elephant is killed before removing the tusks. Hopefully knowing this fact will reduce demand for ivory trinkets.

Poaching (not sport hunting) pressure has resulted in an adaptive change in elephant females – "no tusks" provides a sort of immunity to being killed by poachers. Through a quirk in the genetics of tusklessness, males will not be saved. Attention has been given to the possibility that antlers of white-tailed or mule deer will decrease over time. Elephants were slaughtered without the kinds of restrictions that exist for deer harvest, and to date, it's not clear that hunters are having an effect on antler size in deer.

7. 8.7 Mm = 0.00034251968503937 Inches: A Little Lead Goes A Long Ways

Cabin Talk.—Although I started deer hunting at a relatively late age, our family quickly grew to love venison. On probably one night a week, venison is on our menu. Of the many deer I have harvested, they were all with archery equipment. Although I can process deer myself, I prefer to support my local butcher (and I really like his summer sausage). I know that when I bring him a deer, the parts that go into the grind pile (for sausages, brats, etc.) will be mixed with deer from other hunters, many of whom use a gun and lead bullets. The expectation is that when the people at the shop are butchering, they will thoroughly clean around the hole left by a bullet, so as to avoid getting any lead fragments in the meat. Or so I thought.

I took a deer to a famous big game processor in Nebraska and got back some of my favorites, summer sausage, hot dogs, etc. In the ensuring months, I bit into two hot dogs and one piece of summer sausage and had the "oh no" moment, like when I bite on a pellet in a piece of ruffed grouse. In the accompanying photo, it is obvious that these lead fragments were about 1/8 of an inch, not easy to miss.

This was especially annoying because my deer were shot with a bow. I realize that processors pool trim and make their various sausages from many deer, and I have no problem with this. I have been in several butcher shops and see that they do not use anything that even remotely looks tainted. I mean, if you were a butcher, would you put in a piece of tainted meat and spoil the entire batch? Butchers are smart business men and women.

Yet in my case, it is obvious that at least one deer did not have its wound channel sufficiently cut away and these big chunks of lead made it into the grind pile, into my hot dogs and sausage and into my mouth. Bad, very bad processing. If I had to guess, I'd say it was more than one deer that was not properly butchered, maybe it was a newbie behind the knife. Fortunately, I did not require a trip to the dentist, as I did when a piece of shot broke a cusp off of one of my molars.

The chunks of lead I found were pretty big. It is no surprise to hunters that lead bullets fragment and spread into the adjoining muscle tissue surrounding the wound channel. We've known this for a long time, and the even the potential presence of lead has hindered putting hunter-donated venison on food shelves. Most people who process deer make sure the wound channel is enlarged enough to prevent any chunks of lead getting downstream.

But is removing visible chunks of lead like the ones I found enough? A study (in the journal PlosOne) by Adam Leontowich and colleagues from Canadian Light Source and the College of Medicine at the University of Saskatchewan published some pretty frightening results about just how much lead leaves the bullet and enters the meat. These researchers fired lead bullets into blocks of "ballistic gelatin," the stuff used by law enforcement agencies trying to identify the gun used to fire the bullet, usually in a criminal investigation. They used ultra-sensitive synchrotron imaging to search the gelatin for lead fragments that came off the bullet. This method is able to detect far smaller fragments than standard medical X-ray imaging.

Leontowich "wasn't surprised that bullets can produce hundreds of lead fragments." Well, ok for him, but the number is depressing to me (even as a bow hunter, my meat will be mixed with hunters using lead bullets). What was most surprising to Leontowich was that some of the lead fragments were the size of a human blood cell, which is 8.7 μm, or 0.00034251968503937 inches, hence my title.

Now it seems unlikely that a single lead fragment the size of a human blood cell would be toxic to a human.

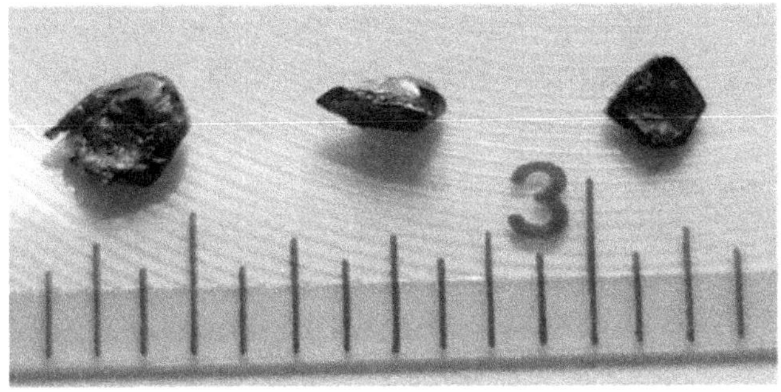

Pieces of lead from venison hot dogs and summer sausage. The author wonders how many smaller ones he missed.

However, given that you can't see fragments this size, you won't know how many you've ingested. It doesn't take very much lead to reach toxic levels. You'd be doing yourself and the scavengers that ingest pieces of lead big favors to switch to non-lead bullets.

And in case you're wondering, I did not go back to that famous processor.

8. Blessed Bucks: Many Questions, Few Answers

Cabin Talk.—Please take this for what it is, tongue-in-cheek satire.

I'm trying to understand the notion of being a blessed deer hunter. I go out hunting and I am hoping to shoot a big buck, one I can post a picture of on social media and exclaim "I've been blessed." There are two clear-cut outcomes, I shoot a big buck or I don't shoot a big buck, so I'll either be blessed or not blessed. Or there's a third outcome, I shoot one and can't find it, that's a category, almost Satanic, well to the south of not blessed.

Only a small percentage of hunters will shoot a big buck, because there aren't very many big bucks. Not everyone can be blessed or there would be no more big bucks. But why was I not blessed and someone else was? Are they a better person? Does the blessed hunter have better skills or hunting land? Are really bad, unskilled hunters ever blessed? What role does luck play? Is there something about my scent? Am I being scent (pun intended) a message that I'm not particularly bless-able? So many questions.

In some views of life, one has to assume that all things happen for a reason, part of the supreme plan for our lives. A big buck apparently is not part of my plan. Is there somewhere I can appeal? Is there a checklist of things I can do to jolt me into the blessable buck category? Is there a random number "blessing generator" and my number just hasn't come up? Maybe I'm in the blessing-in-waiting queue. So few answers.

Of course, the default in the blessed categories might be that if you get to go out hunting, whether you harvest anything or not, just being outside by itself is a blessing. That seems kind of watered down as far as blessings go. Kinda like getting the last Big Mac off the grill at closing time. Or your favorite baseball team being in last place but winning a game. Who wouldn't rather say "Look at this monster I shot" instead of "What a great morning to be outside"?

There is of course an alternative. It's called hunting. That blessed guy was just in the right place at the right time for no special reason, and the deer happened to be there. Chance can play a huge role in our lives! Could have happened to me, but it didn't. Just like someone will win the lottery, but I can safely predict that it's not gonna be you. So, it is perfectly ok to congratulate the hunter who shows off his or her blessed buck, while secretly despising them and wondering why it wasn't you?

9. Flu, COVID-19, You and Deer

Cabin talk.—Every surface of the human body is covered with bacteria, viruses, fungi and other little "cooties." The viruses in your body, your "virome," can number close to 400 trillion individual viruses. And we like to think we are the rulers of our castle. The creed of a successful virus is simple: mutate faster than the host immune system can respond. But viruses, like every other organism, are in perpetual arms races with other organisms – a virus gets better at infecting, the host evolves a better immune response, and on it goes, round and round.

The Human Condition.—We understand exactly how the annual influenza undergoes mutation and as a result can evade the human immune system and our vaccines. So, even though you were immunized against the flu in one year, ongoing mutations make next year's flu sufficiently different genetically to prevent the previous immunization from working. This is not a failing of the vaccines, you won't get an old flu version. Instead, it is an example of an "evolutionary arms race" between your immune system and the virus. The virus acquires new mutations that allow it to succeed, and your immune system puts up new roadblocks, and the process escalates each year.

We are not alone in the flu world. Pigs, birds, cattle, horses, among others carry some form of an influenza virus. Wild birds carry both low- and highly pathogenic influenza, which rarely kills them. However, when a high-pathogenic flu virus such as the deadly H5N1, or bird flu, gets into captive flocks (e.g., chickens, turkeys, ducks), the results are devastating. A few years back, millions of chickens were killed in Europe and Asia to prevent further spread. Health and poultry officials are currently watching closely for the spread of bird flu, which has been found in the US. Why are some flu outbreaks much worse than others? Really nasty flu strains arise when different viruses infect pigs and chickens, trade genetic material, and are passed back to humans. The technical term is antigenic shift, as opposed to antigenic drift, which cause the annual flu viruses to be different. In modern parlance, pigs are where viruses go to become trans-viruses.

Unlike flu strains, only recently have we noticed that the COVID-19 virus has spilled over into other species. The coronavirus mutates rapidly and spread occurs throughout the year. We know a lot about the makeup of the COVID-9 virus. Many mutations in the virus genome, some 29,855 base pairs long (see https://www.ncbi.nlm.nih.gov/nuccore/1916859392 if you want to see them all strung out) are "harmless" but some mutations especially in the "spike" protein allow the virus to escape your immunity, whether provided naturally or via a vaccine. There are thousands of variants but only a few are damaging enough to people to get a popular name

(Omicron, Delta). I suspect the need for maybe two boosters annually, rather than the single one for flu.

The Deer Condition.—As if deer don't already have enough diseases, we have learned that white-tailed deer carry COVID-19. Many other animals have also tested positive (e.g., mink, zoo animals) or have been experimentally infected (cats, dogs, ice, pigs, rabbits, shrews), so this is not major news, except for it being deer. The virus has been reported in deer from at least 15 states and we can expect this number to increase rapidly. In Minnesota, a report from the University of Minnesota found about 6% of deer were carrying the virus, whereas in other states, it is much higher. In some deer populations from Iowa, the prevalence is up to 80%. Two main questions are looming. First, how are deer acquiring COVID-19 from people? Second, will they serve as a reservoir in which the virus can mutate into new strains and then reinfect people with a strain we're unprepared for? Given what we know about flu strains, the latter might be inevitable. That is, deer might act like pigs, in being a platform for different viruses to obtain new mutations they wouldn't have otherwise gotten going in humans.

What is the timeline? Studies of deer tissues taken before 2020 do not show presence of the SARS-CoV-2 virus (they have other viruses). The COVID-19 virus appears in deer samples taken in 2020 and afterwards. So, that answers a common question – we did not find the virus in deer because we just started looking.

How did deer get the virus? It is suspicious that the virus appears in deer at the same time as in humans. But how could people spread coronavirus to deer is perplexing (unless it was the reverse?). There is evidence that infectious viruses from people can be shed in blood, feces, and urine, although the prevalence of virus in these sources is lower than that in the human respiratory tract. How likely is it for someone with an active COVID infection to sneeze on a plant a deer later contacts? People have urinated and defecated (along with bears) in the woods, and this could be the source. Or maybe through water sources – what's next, COVID-19 in walleyes? It might take just a few deer getting it for it to then spread among the herd quickly, especially in winter when deer are in large herds and live in close proximity, with mutual grooming. So, we don't know is the answer to how did deer get the virus.

The virus variants Iowa deer were carrying are of interest. I initially thought they would be deer specific, but the variants (up to 12) that were sequenced from deer match those also found in people in Iowa near where the deer lived. The fact that deer have up to 12 variants means they didn't get it from just one person. It's been transmitted multiple times over a very short period! Deer from Staten Island have been reported to have the Omicron variant! There seems no question that humans are transmitting COVID-19 to deer. How is still unclear.

If the virus variants found in deer are already in humans, maybe it's not worrisome. However, there's no guarantee that after spending time in deer, the virus will remain unchanged. The virus could, in its new deer home, be subjected to pressures that favor new mutations, ones that could be innocuous to deer but deadly to people. Again, think pigs. It would seem simple for the virus to go from deer to people given our love of venison, suggesting heightened vigilance and deer monitoring, but that's unproven.

Are COVID-19 deer getting sick? Unlike mink infected with COVID-19, there is no evidence of sickness in wild deer as yet. Deer experimentally infected with the

SARS-CoV-2 virus don't show clinical symptoms. An obviously huge and looming question concerns eating an infected deer. No one knows if you can get the virus from a deer. The CDC suggests cooking to a fairly high internal temperature, well past the rare level at which I like my venison.

What to do? We still haven't solved CWD, and now this. Dr. Tony Goldberg (UW Madison, professor of epidemiology) told PBS Wisconsin that an evidence-based approach warrants monitoring the occurrence of virus in deer, determining the extent of its spread (upper Midwest, entire range?), and possible culling to limit spread. So far, the good news is that deer do not get sick from the virus. It would be even better news if they keep their new variants to themselves.

10. Trail Cameras and Deer Hunting

Cabin Talk.—Technology plays a large role in our lives. Much of it betters our lives, like pacemakers and cell phones, although sometimes the latter is debatable (remember newspapers?). If you wanted to make a list of all the ways this is true, it will take you all day — start with your car. Other kinds of technology give us a leg up on things we might like to do, but we don't really need. Take electronics in fishing. I have a gizmo on my boat that finds fish not just underneath my boat but out to the side. Without it, I would easily miss a school of walleyes, one I have missed countless times prior to this technology. Even newer is forward-facing sonar, which provides real time images of fish, and much debate is ongoing about whether it's fair. And then there are trail cameras, useful and informative in their own rights, but also providing hunters valuable intel on where and when deer frequent a given spot. Where, or do, we draw the line?

When trail cameras came out, I followed the herd. What a pain it was to load the 35 mm film, go check the camera, decide if there were enough pictures snapped to take to the drug store to get developed. Then home I'd go with a bunch of grainy photos of rabbits, racoons, squirrels and some deer. I then jumped at the chance to get the trail cameras that used the early versions of SDHC cards, so tramping out to the camera and trading out the card was a huge advance over the two trips to the drug store. Still, it was a trip to the camera, all the while leaving scent everywhere, and then either downloading the images off the card on the spot with a card reader or bringing it home to read on my computer.

Now, I've succumbed to the cellular trail camera, which notifies the app on my phone when it takes a picture. My wife is unimpressed when I forget to turn the phone off during the night and my phone vibrates every time a racoon, opossum or deer triggers it (doesn't wake me with my CPAP machine running).

What has all this intel done for me? I'm not an antler hunter, so I am more thrilled to see a doe on the trail cam than a monster P&Y buck. I'd try to pattern the deer on my camera and time my sits to maximize my chances of bagging that nice doe. But it didn't work for me. Deer don't get the memo that says, "Do the same thing at the same time for at least a week" or "Bob will be waiting at 4:57 p.m. for you to come by, don't disappoint". I found my success was more or less random. It was fun looking at images on the camera. Once in a while I'd see a big buck, but given their enormous home ranges in the rut, it was usually once, when obviously I wasn't there.

There's a move in some western states to ban trail cameras during the deer hunting season. For example, if you hunt deer in Utah, you cannot use a trail camera for five months a year corresponding to the deer season. Rather hilariously, the regulations say that private landowners can use trail cameras to monitory their property provided the cameras aren't used for any type of big game hunting (don't ask, don't tell?). The Utah DWR asked hunters for their opinions, and about half said there should be some regulation of trail cameras, 38% said no to any regulation, 11% were neutral; so it was 50:50. The Republican sponsor said there was overwhelming support for regulating trail cams -- no offense, but apparently, he was not elected for his understanding of arithmetic.

Support for this goes beyond Utah. The Boone and Crockett Club said that trail cameras give hunters an unfair advantage, and it will no longer accept any entries for big game animals harvested with the "aid of transmitting trail cameras and other technological devices, such as drones, smart-scopes, electronic game calls, and even cell phones." I had to look up "smart-scope." B&C says "these technologies can displace a hunter's skills to the point of taking unfair advantage of the game. " It goes against fair chase, some say. As long as they don't have a listing for does, I'm not very worried about their stance.

Well, I do know of some outfitters that pattern bull elk and put hunters in the right spot. I might be in favor of a regulation banning use by outfitters, as the hunter him or herself should at least do some of the leg work. The Pope and Young Club has refined their objection to cellular trail cameras, to when the real time images allow someone to dash out and within minutes of getting the image, shoot the animal. Also, in the west, animals congregate at water holes and some say that trail cameras are really unfair to animals coming to drink. But who wouldn't figure out to go sit water holes even without camera-aided intel?

For me, the principle issue is whether this has an effect on the deer herd. I took a look at the impact of trail cameras on the Minnesota deer harvest. Trail cameras became prevalent in the mid-1990s to early 2000s. I plotted

the total deer harvest, buck harvest only, and most importantly, success rate at harvesting bucks from 1989 to 2020.

In the accompanying image, you can see that the total harvest and buck-only harvest numbers run pretty parallel. But the number harvested varies over time because of the number of deer, and number of hunters, and changing regulations. Many hunters remember 2002 to 2004 when there were lots of deer and liberal limits (to reduce the herd to prevent greater ecological damage and disease transmission).

The most important observation to me is the relative success rate in harvesting bucks (the dashed line), which ranged from about 25% to 47%. Opponents say that trail cameras are primarily used to target large bucks. The data, however, do not support this idea. I show with the arrow on the graph about when trail cameras became common. If trail cams allowed greater, selective harvest of big bucks, the success rate should have gone up and up and up. It did anything but that. Sure, maybe someone got a large buck by snooping on an individual buck, but if that person didn't harvest it, her neighbor did. I am not convinced that limiting trail cameras is the right thing to do, at least at this point.

11. What Lurks In Your Deer's Genes Might Surprise You

Cabin Talk.—During my lifetime molecular geneticists have learned a lot about the genes that made you you and me me. We don't know everything, but we know that our DNA has some "junk," and some genes or gene parts from a process called "horizontal gene transfer," where viruses in the past were nice enough to share with us a gene or two of theirs. We know that most of our appearance is governed by the genes we inherited from our parents, which is why we resemble them more than the postal person. But we also know that there are agents that influence not the letters of our DNA code, but how the genes are expressed when we develop, that are not part of our DNA per se. BTW, did you know that as an embryo you had a tail and gill slits?

Descent with modification, evolution for short, is the scientific explanation for the diversity of life we see on earth. But modification of an animal's appearance (or a plants) during evolution is accomplished in several ways, not simply by adding one new characteristic and deleting the previous one. In fact, relatively few genes make up the building blocks of an animal, and one of the main reasons for why animal species appear differently is in the rules for assembling the common building blocks, not the blocks themselves. Think of all the things you could build with a set of Legos – enormously different looking things made of the same building blocks; animals aren't that different. Development of a human from a single fertilized egg cell occurs because of instructions encoded in our DNA that amounts to "some assembly required," in this case a colossal understatement.

With ongoing changes in a lineages genome, one might ask whether a structure that was present in your ancestor but is no longer present in you has also been lost from your DNA. Short answer is not necessarily. Genes that have not been used for a long time tend to decay, so-called molecular fossilization, whereas others can be repurposed for other tasks. For example, the huge "horns" in rhinoceros beetles are made from genes that previously made leg segments! (they still have genes for legs, just some were "co-opted") Sometimes genes become duplicated and each "daughter" gene takes on a different function. For example, humans have several forms of hemoglobin, each evolved from an ancestral globin gene, one of which, fetal hemoglobin, shuts off within a few months after birth, never to be used again. But it's still in your DNA, and in theory could be turned back on. Traits that have been missing in an animal species but reappear in a few individuals are called atavisms.

I read about a buck harvested in Louisiana in 2016. The hunter (Wilbert Collins Jr.) noted that the buck had distinctive markings on the top of its head unlike those we see in typical white-tails. These head markings consisted of a black line up the snout, which branched at the eyes and continued to the base of the antlers. In other words, it formed a "Y" with the bottom of the Y at the nose.

The strangest characteristic of Collins' deer was noticed when the skull was cleaned. It had fangs (see photo), which are actually upper canines. If you do an online search for white-tailed deer with fangs, you'll find images of fanged white-tails and mule deer from several states. C. W. Severinghaus reported 23 instances of upper canines in about 18,000 (0.13%) white-tails from New York. Two other researchers, Knowlton and Glazener, reported a much higher frequency of 17.9% in white-tailed deer from the property of the Welder Wildlife Foundation (Sinton, TX). However, they noted that a large percentage of upper canines did not protrude through the gum and were unnoticed until after the skulls were cleaned. Some have suggested that upper canines are more prevalent in southern deer. However, given that skulls are often not cleaned, the actual frequencies of upper canines in northern deer are unknown – I'll be rubbing my finger over that spot from now on.

Fangs in deer are nothing new. As many readers know, muntjac deer of southern Asia and Chinese water deer have well-formed fangs, which function like tusks, and are used by males fighting for dominance and access to females. Fangs were present in ancestral deer. For example, a now-extinct deer occurred in Nebraska about 12,000,000 years ago at the famous Ashfall fossil beds (http://ashfall.unl.edu/ashfallstory.html) - it stood 2 feet tall at the shoulder and males had muntjac-like fangs and lacked antlers. Maybe if you're a two-foot tall deer, the fangs acted like a predator deterrent? Evolutionarily these fanged deer are distantly related to white-tails, meaning

that the traits could have been independently acquired, that is genetically engineered in two different pathways, or genes that were silenced are now activated but are the same old genes.

The pattern of head markings in the Collin's deer resembles those found in the muntjac deer, although the Collins' deer's markings were somewhat different. A noted Canadian specialist on hooved mammals, Valerius Geist, pointed out that white-tails have a black stripe on the lower jaw exactly where the upper canines would protrude from a muntjac, or an ancestral deer, and he suggested that

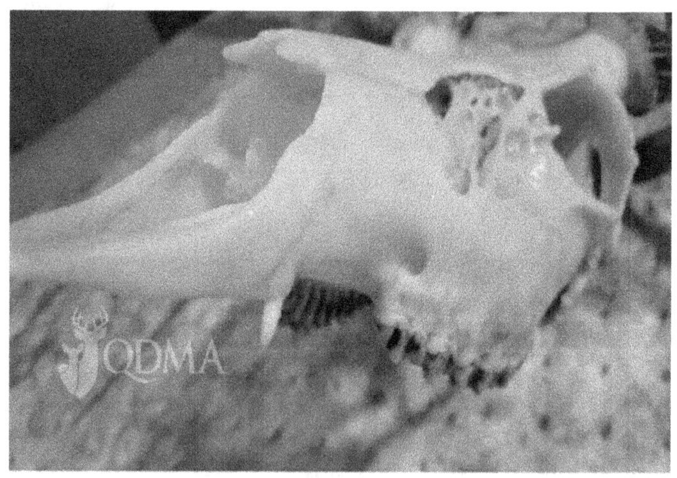

Prominent "fang" in white-tailed buck from Louisiana. Photo courtesy of Wilbert Collins Jr. and QDMA.

in bygone days it would accentuate the fangs appearance! Why else would that black stripe be there, probably has, or had, some function? Go look at one of your mounts.

How did these head-markings and fangs of the Collin's white-tailed deer come about? Because these traits are found in the relatively primitive muntjac and Chinese water deer, but not in other deer, the most likely explanation is that these are traits of ancestral deer, and fangs were replaced by antlers as the favored tool that bucks use to assert dominance and attract mates. Furthermore, these traits are probably genetically linked.

The fact that we sometimes see upper canines in white-tailed deer and mule deer suggests that the genetic architecture remains and involves a relatively simple switch that occasionally gets turned on, causing the fangs to appear. It will be easy to find these silenced genes with modern DNA technology.

In the genetic makeup of all animals lurk many silenced genes, and they occasionally reappear. For example, embryonic humans have a tail between days 31 and 35, and sometimes, the tail persists after birth. In humans a condition called werewolf syndrome (hypertrichosis) causes hair to grow over the entire face, although it is very rare. When we compare human and chimp genomes, the similarity is about 94%, which seems high for how different we look. However, the high genetic similarity is not the whole story. For example, humans are missing part of a gene that in chimps tells the brain to stop growing, whereas in humans that gene is "silenced" and it results in greater brain size. So, the rules for putting together a human or a chimp are very different but involved most of the same building blocks.

So, finding what appears to be bizarre features in deer or any other animal is usually a result of the expression of a typically silenced gene that somehow gets put back in the developmental circuitry. These atavisms are very important in teaching us about what animals used to look like and why they look like they do today.

12. One Tough Deer

Cabin Talk.—If you watch things in nature long enough, you're bound to see something out of the ordinary. Common "rare" sightings include albino deer and robins, woodpeckers with very long beaks, or a snake with two heads. Sometimes animals do things to other animals that catch our attention. A bird at your feeder without a tail likely barely escaped a hawk attack. Tails on birds do more than help with steering, they can come off if a predator attacks, allowing the bird to escape. Sometimes a hostile interaction between two animals leaves actual, visual evidence.

There are many images on social media of bucks whose antlers locked during battle and either one or both

A. Wounded young buck in Woodbury Minnesota

perished —the pursuit of love can be costly. Even more common are bucks who have broken tines or might be missing all or part of an entire side from past fights; incidentally, this is why it's a good reason to regrow antlers annually. It is common to see bucks with some nasty wounds on their bodies obtained in fights for dominance. However, what follows is an account of a fight between

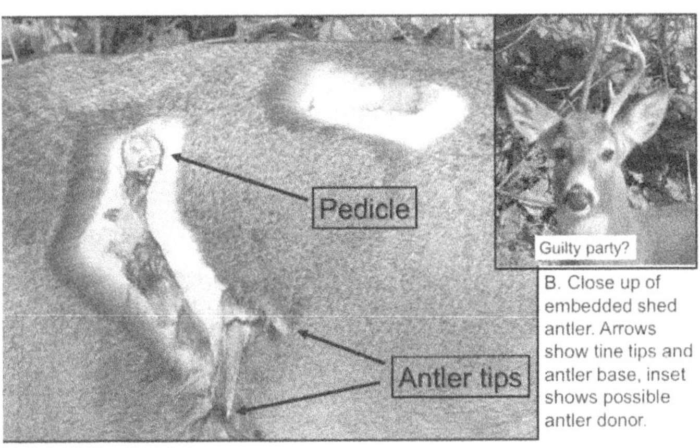

B. Close up of embedded shed antler. Arrows show tine tips and antler base, inset shows possible antler donor.

two young bucks that left visible, almost unbelievable evidence.

The buck in question (Images A, B) showed up with a few other deer in Woodbury, MN, on March 5, 2021. He was obviously injured and after some conjecture on what might have caused the injuries, there was no consensus. The next day he was seen with a buddy, who was sporting a single antler (Image B). The wounded buck kept coming around for the next few months and the last sighting of him was on May 4, 2021, and surprisingly, he was well along in growing new antlers (Image C).

A closer look at Image B revealed that the wound was more than a puncture. This buck had another bucks antler impaled under the hide in his side! You can see the underside of the base, or coronet of the shed antler up by his spine. The tips of at least one and possibly two tines (arrows in Image B) are protruding about midway down the side, and they would seem to match the curvature of an antler main beam, albeit not from a huge buck. Kip Adams, Chief Conservation Officer and deer expert from the National Deer Association commented "I've seen numerous antler points impaled into a deer's face, but I've never seen an entire antler as you have here."

At the risk of over speculating, the antler on the one antlered buck in the photo seems suspiciously similar to what you might outline under the hide in the wounded buck. Maybe these two bucks were doing some fighting and it got heated. The antler could have become embedded after he jabbed the other deer, and when the buck turned away, the antler got caught under the hide and sep-

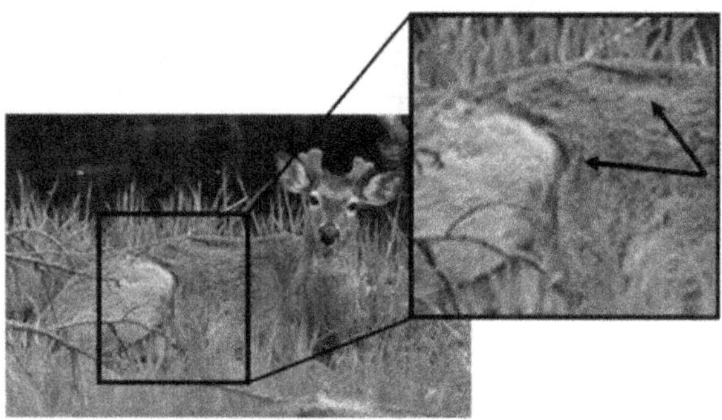

C. Buck healing and growing new antlers. May 2021

arated, because it was soon to be shed anyway. It must have somehow worked its way down the side and broke through the skin. Again, we are speculating, as the evidence is circumstantial at best. As a caveat, we suspect that the embedded antler is from the left side, whereas the one antlered buck in the image still has his left antler, which throws a wrench into a guilty verdict for him, but we'd like to see his brother!

The hair around the wound appears almost as if it had been shaved, which is probably a result of licking, as suggested to us by Lindsay Thomas (National Deer As-soc.). The wound on the top of the back was probably a separate puncture that was also licked by the deer.

Although, we do not know the ultimate fate of the wounded deer, one might guess that infection would take its toll. However, in Image C you can see that our wounded buck is growing new antlers in May! This is a reminder of the resilience of deer - they can soldier on after some major injuries and then heal quickly.

13. Why Aren't Deer Like Rabbits?

Cabin Talk.—Sometimes it's fun to just sit and drink some coffee and wonder. Wonder about ideas you might not want to share with anyone, lest you venture into the dumb question zone. But it's fair, and ok to me, to do some "what if…" type of thinking. How come things are the way they are, and not different? I had one such vision, pun intended, that resulted in thinking about coloration of deer and rabbits.

Deer are like rabbits, at least in some ways. Both are primarily plant eaters, although deer will eat eggs and baby birds if they find them. And their "calling cards" or droppings can sometimes look pretty similar. Unlike rabbits, however, deer are not coprophagic (a good Google exercise if you're unfamiliar with the term). They are both sedentary, meaning they don't migrate to escape hostile winter conditions and food scarcity. Northerly populations of both species have survived multiple cycles of glacial advances and retreats, and currently live in areas that 20,000 years ago were covered by a mile thick glacier.

That is, 120,000 yrs or thereabouts they were driven southward by advancing glaciers, survived all of these years south of the ice sheets wherever there was suitable habitat, and then followed the retreating glaciers and regenerating ecosystems as they recolonized northward. Both have survived for tens of thousands of years in cold winter environments. They must be pretty well adapted to winter conditions, although harsh winters can result in population declines, but they recover.

I saw an image of a nearly albino white tailed deer, or at least a piebald, and it got me thinking. Obviously, it is within the potential of a deer's genetics to have white fur. Even normal deer have white hair on their throats, ears and bellies, and there are white spots on fawns (carried into adulthood in some species like axis deer). Everyone knows that snowshoe hares turn white in the winter. The obvious reason is that it is a way to escape predators by matching their winter white environment, called background matching by biologists. With spring and new vegetation, they molt back to the brown pelage, which functions as summer camouflage. Not just rabbits pull off a seasonal coat change. For example, many birds have distinct alternating winter and breeding season plumages.

The importance of background matching is clear. In the case of snowshoe hares, studies in Wisconsin have shown that with warmer winters, the distribution of hares has shifted northwards to coincide with the southern limit of the snow line. The reason seems clear enough. Hares that stayed at the former southern range limits are exposed to a hair/background mismatch at least in early spring because today snowmelt is earlier; yet hares continue to molt each fall into the white coat. As you might expect, a white hare against a brown background is easy prey. Why haven't the hares just started to molt earlier in spring? The twice annual molt in hares is almost certainly coupled to photoperiod, aka day length. Whereas the winter snow line has shifted northwards, daylength has not, and hares had to shift their range northward because the molt to white coats is locked in, for now at least, and the trigger is day length and not snow cover.

Why don't northern deer change their coat colors with the seasons? It might be adaptive to grow a white coat in the winter. Perhaps the difference between deer and rabbits is about predator avoidance. Snowshoe hares are hunted regularly by goshawks, the largest of North America's bird-hunting hawks. Sharp eyesight allows goshawks to pick off movement and a brown hare against a white background probably makes the goshawks job way easier.

Deer on the other hand have co-evolved for millennia with predators such as cougars and wolves, both of which are likely to use their sense of smell instead of sight to find their prey. Thus, perhaps a white deer standing in the snow is just as visible to a wolf as a brown one. Maybe the best predator avoidance strategy for a deer is to evolve keen hearing, night vision, and as little odor as possible (as apparently is the case for fawns). On the other hand, if a deer was white and upwind of a predator, it might avoid visual detection. In addition, there might be a thermoregulatory reason, in which a brown deer might be warmed more readily than a white one on a sunny winter day.

I wondered if albino deer live longer. I don't know of scientific evidence but the popular opinion is no. They, like hares, are badly mismatched when there is no snow on the ground, and albinos don't molt into brown deer in summer, they're white 24/7/365.

Could science come to the rescue? Researchers have done winter season transplant experiments introducing snowshoe hares with white fur into winter environments without snow; it doesn't turn out well for the hares.

At least half seriously, the obvious experiment would be to paint a bunch of white tails white, radio collar them, and see if their survival was enhanced in winter. Of course, you'd need to do scientific "controls," by capturing deer and painting them with water instead of white paint, to see if the mere trauma of being captured and painted made them behave in some unusual way that affected the probability of capture by a predator. I realize this would not be a trivial undertaking!

As far as I know, there are no deer species that alternate seasonally between white and brown fur. A natural experiment would entail a mutation or two, like in hares, that causes the later summer molt to result in white fur, and conversely back to brown in the late winter. Actually, the question is whether this hasn't happened because the right mutations haven't come along, or because it would not be an advantage to deer like it is to hares.

14. It's Taken Centuries, But We Now Know Why Deer Don't Ask To Use Your Compass

Cabin Talk.—Natural history is about learning basic facts about plants and animals, where they occur, what they do, how they interact, and so on. Many scientific journals are devoted to reporting observations that scientists make about natural history. The public, however, sometimes does not fully understand what "natural history" is about, as the editor of one prominent journal reported that he received a letter asking for a list of nude beaches in a Latin America country. The topic here is about something that is incredibly fundamental to many animals, that we have simply not noticed, for centuries.

At one time I would have wagered that we have missed relatively few "obvious" things that occur in nature. We know about migration, hibernation, what habitat birds use, who eats whom, who's the fastest, etc. But not all things can be observed with the unaided eye, and some things that animals do require sophisticated technology for us to see or understand.

For example, we have learned that many birds have a well-developed ability to see in the ultraviolet part of the light spectrum. And with scientific experiments, it is now well established that birds communicate visually with patches of UV plumage that we simply cannot see — we miss much of what birds are telling each other because we are blind to the signals.

Still, you'd think that something basic, something that could be observed with the unaided eye, would not have escaped centuries of observations. Right? Apparently, we have missed something. It was reported by Sabine Begall (from Essen, Germany) and colleagues Jaroslav Červený, Julia Neef, Oldřich Vojtěch and Hynek Burda, in one of the most prestigious scientific journals, the Proceedings of the National Academy of Sciences of the USA. But before I give it away, here's some background.

Ranchers have known for a long time that most sheep and cattle in a group face the same way when grazing. Farmers know that cattle usually face into the wind, whereas sheep face away from it. On a cold windy winter's day, you'll most likely see cows (and hence the whole herd) orient themselves into to a strong wind, exposing the least amount of body surface area. When it is cold and the sun is out, you'll probably see grazing animals orient perpendicular to the sun to absorb the most solar radiation.

These behaviors occur in stressful conditions. But what happens when conditions are not particularly stressful? One would think that the direction of body orientation would be random. Turns out that cows, and two kinds of deer in Europe (red, roe) orient themselves very strongly in a north-south direction in non-stressful conditions! No way, how could we have missed this?

How Begall and her colleagues figured this out is pretty cool. Believe it or not, they looked at maps on Google Earth and that they could see cows in pastures and determine which direction they were facing. I was skeptical, so I asked Begall to send me coordinates so I could see for myself – they weren't kidding!

I tried it myself by visiting Google Earth and entering 36.58'57.57" S, 61 42' 11.55" W. Sure enough I saw a herd of cows in an Argentine pasture oriented N S, from space! They actually used images like this to figure out which way the cattle were standing! (Note, this won't work today because Google Earth updates images).

Begall and colleagues reasoned that because their images were from Africa, Asia, Australia, Europe, North America and South America, and Australia that it was unlikely that the wind and sun were the same in all places, and that these could not explain the consistent N-S body orientation. They concluded that cattle were using the earth's magnetic field. We do know that many animals, like birds, use magnetic lines of force to navigate. It now appears that cattle and deer also can sense magnetic lines of force and use them to orient themselves.

But wait, what about deer? Well, they have field observations (not satellite) of roe deer and red deer in the Czech Republic. They found that both resting and grazing deer tend strongly to orient in a N S direction.

Even more impressive, they went out and found deer beds in the snow, finding again a very strong N-S orientation! They also concluded that for resting roe deer, not only was the orientation N-S but that the head of the animal is almost always pointed north. Somewhat comforting to the natural historian in me, they observed that in herds of grazing red deer, about one third of the animals were oriented south, which they figured was an anti-predator behavior. So, it's not all or none.

Their observations certainly raise some questions. I would think that resting deer would orient into the wind to detect approaching predators. However, the authors concluded that at least at night, bedded roe deer and red deer were deep in forests where wind was damped, where hearing becomes the main defense.

The obvious big question is, why? The authors also recognized this but could only make some suggestions. There is safety in a herd because of many eyes and noses. Perhaps maintaining a certain magnetic orientation might provide a directional reference for the animals in case the herd is scattered, and everyone moving in the same, not random, direction helps the herd to regroup as quickly as possible.

The authors point out that the phenomenon requires more study, but one cannot help being awed by their ending comment: "It is amazing that this ubiquitous conspicuous phenomenon apparently has remained unnoticed by herdsmen and hunters for thousands of years."

Now when you see whitetails feeding in non-stressful conditions, or in the winter when you find beds, record whether they are indeed oriented in a north-south direction. Remember, it's not likely all-or-none, but it would be interesting to know if North American deer also show a strong tendency to orient N-S. But there's a potential catch. The researchers found that the direction of orientation followed the magnetic north, which can be different from simple geographic north.

15. Deer And Optometrists

Cabin Talk.—You're out hunting, creeping silently through the forest, keeping your movements slow, deliberate and to a minimum. There's a decent buck just around a large oak tree, head down, feeding away, oblivious to your approach. It's the last day of the season and the unused buck tag calls from your backpack. Just a step away from a perfect shot opportunity, the buck looks up as just your foot is contacting the ground. Movement! Poof, stalk over, deer gone.

So how is it that last year driving to the cabin on a Friday night after work, your car lights blazing, engine revving, moving at 65 mph, you hit a deer on the highway? Where was that "I see you making the slightest of movements and I'm gone" response then? Your vehicle is a lethal weapon as well and you'd have to think that every deer family has lost at least one loved one to a vehicle. How can we talk about how well adapted animals (and plants) are when a deer is not smart enough to get out of the way of an oncoming car? And what's more, why do they sometimes start to run across the road and then STOP on the road, almost like volunteering to be fodder at the wolf center?

Short answer: all adaptations are compromises. I'm reminded of a quote from author Christopher McDougal: "Every morning in Africa, a gazelle wakes up, it knows it must outrun the fastest lion or it will be killed. Every morning in Africa, a lion wakes up. It knows it must run faster than the slowest gazelle, or it will starve." All animals are in a perpetual "arms race" with their prey or their

predators or both. Natural selection has an extremely good ability to lead to faster and faster gazelles, but so too does it lead to lions with greater and greater powers of acceleration.

But all machines, whether built by man or evolved via natural selection, have limits or constraints. To run faster, gazelles over time might evolve longer legs allowing greater speed. But legs that are too long will likely break much easier, and it doesn't matter how fast you could run if your leg is broken. Also, there's the mechanics: as a leg

Eye shine in white-tailed deer caught on a trail camera at night.

gets longer, after a point they also have to be wider and heavier so as not to fracture, and then the increased weight negates any increase in speed owing to length (and it takes more energy to grow and maintain heavier bones). It's an engineering problem, and the result is that an adaptation for one thing in your life might hold you back in others.

Animal bodies are full of compromises between competing needs.

Take the peacock. The long train of feathers that insert on his back are beautiful to us and peahens, but at some point, it won't matter to the male in terms of getting more and more mates if he's eaten by an average predator because he's too clumsy to take off in time. BTW, can't help reminding, the feathers we all admire on peacocks are not part of its tail, which actually consists of very average dull colored feathers.

Back to deer on roads. Deer have 20/200 vision, whereas people often are 20/20 (before they start reading and staring at screens), meaning that a person can discern details at 200 yards, but a deer would need to be 20 yards away. Deer are good, however, at detecting motion.

Deer are crepuscular and nocturnal creatures. There's a reason many deer are harvested just before dark or just after sunrise because that's when they're most active. And this activity pattern is not because they were too lazy to get up earlier, even for deer in their teens. Their eyes are best adapted to low light conditions, unlike yours and mine.

If you've been to an optometrist in the old days, you've probably had your eyes dilated when they put some drops in them (such as Tropicamide, a mydriatic drug). This makes the pupil much larger (reminder, the pupil is the black spot in the middle, not the colored portion that is the iris), allowing the doctor to see to the back of your eye and examine your lens, retina and the fluid in the eye. By opening, or dilating the pupil, more light is let in. If the clinic suddenly became pitch black, you'd have a better chance of seeing than the optometrist.

But then, when you're ready to leave and the effects of the drug have not worn off, your eyes are very sensitive to light because of that big old hole in your eye, and without your own or the very stylish plastic rolled up sunglasses they give you, you'd probably have an accident while driving home.

Same with deer on the highway at night, their pupils are wide open to gather light so they can see. But when a bright light shines in their eyes, like the light from your headlights, they are blinded. This triggers the second not so adaptive in this situation adaptation, the fear response. That is to freeze and not draw attention to yourself.

Yes, it's every animal for him or herself. If a lion surprises you and a bunch of other gazelles, if you freeze and another one starts to run, the lion is likely to attack the one that is moving. So, in the case of an oncoming car the two adaptations lead to a bad situation for the deer, driver, and insurance company, but a positive one for the car body shop. I did mention that life is a compromise.

So, how did deer get to this point? To adapt to seeing in low light, differences evolved in the eye's structure. The lens in a deer eye is larger than that in a person, allowing more light to penetrate to the retina (the screen in the back of the eye). Our lens' are slightly yellow, allowing them to filter out damaging UV radiation encountered during daylight. Lacking this yellow pigment, deer lens are clear and in fact allow them to see in the UV spectrum, further enhancing night vision.

Vertebrate eyes have two types of photoreceptors on the retina, rods and cones, the former help with distinguishing light and dark, the latter with color. Deer have more rods that humans, allowing deer greater night vision capabilities. Their color vision is weak relative to people, with deer seeing red and orange poorly, but they are better at seeing blue and yellow.

The human pupil is round and when dilated doesn't expose most of the orbit. However, the eye of a deer is elliptical like a cat, and when it's dilated can (un)cover most of the eye. Likely if our eyes were like those of a deer, the optometrist wouldn't let us leave the clinic until the drug wore off.

Lastly, what about that "eye shine" that many animals, not just deer, have? The shine is a result of light reflecting off a structure behind the retina called the tapetum. It has a doubling effect, by reflecting light back again across the retina, allowing better low light vision. If, however, the light is too bright, it has a blinding effect, fatally so for a deer crossing a highway.

In the end, an adaption to one life aspect results in a disability in another. Obviously, deer didn't evolve in an environment that included blinding headlights at night. The larger, more dilated pupils, larger lens, high proportion of rods, and the doubling effect of the tapetum leads to more light reaching the retina, allowing deer excellent night vision. However, bright lights from headlights overwhelm these same adaptations leading to "oversaturation," or temporary blindness, and potentially being clobbered by your car, a decidedly non-adaptive result.

16. Does Your Deer Have Keds?

Cabin Talk.—All animals have parasites and many parasites have parasites. Deer are hunted and watched extensively and have a number of easily visible conditions traceable to parasites. One of these parasites is the deer ked. Never heard of it? Neither had I until recently. An editor thought it was kids shoe. My spell checker wanted to change it to something else!

Deer keds are highly specialized flies, also called louse flies, or for good reason, flat flies. Four species occur in the US, although one of them is introduced. They have a fascinating life history. Deer keds are related to similar flies on birds. When I have seen them while handling recently deceased birds, they are almost like ghosts, as they move laterally through the plumage, and trying to catch them is really hard. Now you see one, now you don't, is usually how it goes. This behavior allows them to avoid being picked off by the bird too.

These insects are obligate parasites that feed on the blood of their hosts. The larvae develop inside the female and are fed a special secretion internally. When mature, they are deposited in the soil and immediately pupate (form a cocoon), where they stay in soil or leaf litter for almost a year. The pupae basically fall off a deer, so they are most common in bedding areas.

Although you might think that keds find a host when a deer rolls over on them, in fact when the larva emerges from its pupa on a warm day (usually fall), it flies to another host. Unlike mosquitoes, which use olfactory or scent cues like CO_2 to find their next meal, these flat flies are attracted to movement, typically a large mammal like a deer. There are, however, reports of these flies landing on and biting people. After landing on an animal, they drop their wings! The wingless keds crawl around the deer and can crawl from a deer to a hunter or butcher.

Although deer harboring flat flies do not apparently show impacts, the same is not true for moose, where deer keds also feed (perhaps they should be called deer-moose keds). In Europe, these parasites can be so abundant on moose that it leads to partial or complete hair loss (alopecia), which can be fatal in hard winters. The species in eastern US was accidentally introduced in the early

1900s and they have spread throughout Pennsylvania, but only sporadically elsewhere, which seems almost odd because deer are continuously distributed and I'd think provide an avenue for ked dispersal. Given the issues midwestern moose have with parasites like winter ticks, hope-

Top: European deer ked in winged stage. Photograph by ML Legrand via iNaturalist, used under a CC BY-NC 4.0 Deed license. Bottom: European deer ked that has dropped its wings. This is the form that would be seen on a deer. Photograph by mistycal via iNaturalist, used under a CC BY-NC 4.0 Deed license.

fully keds stay put, our moose don't need any new challenges. Hopefully, there's some biological factor(s) preventing westward spread, although the species in the west also present a potential threat.

As far as people are concerned, about 9% of deer keds in Pennsylvania carry *Anaplasma*, which can cause a potentially fatal disease, granulocytic anaplasmosis. Although it is not clear anaplasmosis can be transmitted via ked bites, the Penn State extension folks suggest treating a bite from a ked like you would a tick, and seeking medical attention should you develop flu-like symptoms after being bitten. Interestingly, common tick repellants like DEET have no effect on deer keds; however, it is reported that permethrin kills keds 100% of the time. Of course, you're not supposed to treat your skin directly with

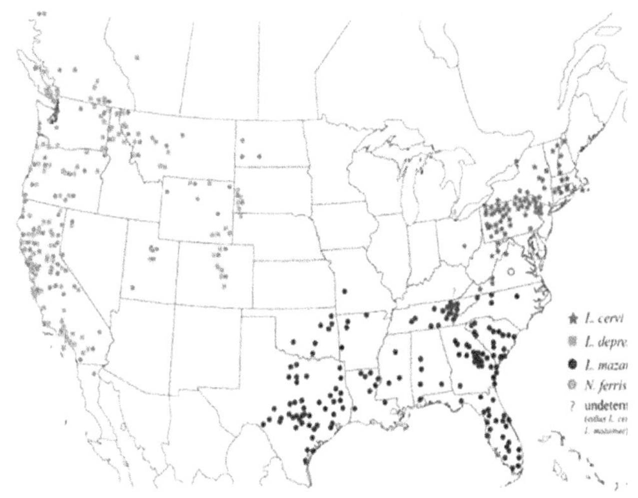

Distribution of deer keds in North America (https://extension.psu.edu/deer-keds). *L. cervi* is introduced from Europe.

permethrin, so if it was of concern, deer skinning probably should be done with gloved hands. Lastly, studies in Europe suggest that light-colored clothing is less attractive to deer keds than dark-colored clothes, which makes sense if they're searching for deer or moose.

The world of parasites is a fascinating one. The most common surface for an animal to live on is another animal – most animals have parasites (humans have around 300) and many parasites have parasites. This is because parasitism is evolutionarily very successful. Deer keds are examples of a "prudent" parasite because it does not kill its host, rather it mitigates the harm it does so that the host population survives.

17. Basics of Chronic Wasting Disease

Cabin Talk.—The annual harvest of white-tailed deer and mule deer in the US averages around 6,000,000; Texas probably leads the harvest with around 450,000 taken by hunters each year. Over 10 million hunters pursue deer. Deer hunting contributes billions of dollars to the US economy. Given the social and economic aspects of deer hunting, Chronic Wasting Disease has the potential to make a huge negative impact. Even if the deer herds are reduced by 50%, the consequences would be enormous. Below I provide basics of the disease, with apologies to those for which this is old news. Other aspects are dealt with in different essays.

Chronic Wasting Disease (CWD) is a 100% fatal neurodegenerative disease in cervids for which there is currently no vaccine available for wild deer. The disease is one of a family of transmissible spongiform encephalopathies (TSE) that includes Mad Cow disease and human versions such as variant Creutzfeldt Jakob disease and kuru. (FYI, encephalopathy is pronounced "uhn·seh·fuh·**laa**·puh·thee" with the accent on the "laa" – I once mispronounced it as encephalo-pathy, and was immediately corrected by a vet in the audience; but hey, I'm a bird guy). In a TSE, the infectious agent is a naturally occurring prion protein that misfolds and becomes infectious, builds up and clogs the central nervous system, causing death. The time to fatality varies, but probably once initiated, is a few years at most. Because they're not living, prions "reproduce" by bumping into normal proteins, which apparently causes them to misfold, thereby growing the concentration of infectious prions. Hence, CWD is nothing like bacterial or viral diseases, in which the body can develop immunity.

How does CWD spread among deer if it's not like viral or bacterial infections? Transfer of prions can involve rodents, inter utero transfer from mother to fawn, soil contamination, sperm, among others. The most common and most contentious route is through saliva from infected deer deposited on corn or other food that is spread to attract deer, either by hunters or nature lovers; hence, widespread bans on feeding deer. The most insidious transmission route involves infectious prions deposited in the environment that persist long after the infected deer has died. These prions are basically a long lasting environmental toxin, like slow-acting cyanide, which can even be taken up by plants (think deer food). This means that even if deer are rare or even eradicated, infectious prions lurk in the soil for considerable periods and will infect deer that subsequently come to the area.

Determining if a living deer has CWD is difficult until the animal expresses the terminal or clinical symptoms. Excessive drooling, emaciation, staggering, loss of fear are some of the symptoms of the latter stages, although these symptoms are also produced by other diseases. There are no known treatments for CWD, although research is constantly ongoing, so it pays to check in on progress.

What does it take to consider a deer CWD+? Many publications seem to suggest that a deer either has CWD or it does not. Technically this is true. State and federal agencies have different thresholds for the concentration of prions required to classify an individual deer as having CWD. That is, a deer might be CWD positive for one agency but not for another. The common test used to detect CWD probably cannot detect the very early stages of a CWD infection.

In this age of misinformation disseminated by those without expertise, many misconceptions exist. For example, some doubt that CWD is fatal. Not so, every deer in captivity that had CWD died – 100% fatality rate; same for all TSE diseases. This leads some to claim that if CWD is fatal, why aren't we tripping over dead deer in the field? The answer is clear, if a deer makes it to the final stages and dies, it's rapidly consumed by scavengers. Ever seen remains of a dead deer in the field? I have. Before the terminal stages, and before we might recognize that a deer is CWD+, it is taken by opportunistic predators or hunters. It turns out that CWD-positive mule deer are eaten by cougars or hit by cars, and their deaths would not be attributed to CWD. And as I note elsewhere, it is certain that many hunters have eaten venison from deer in the early stages of CWD, when they appear and act normal.

Is CWD a newcomer on the disease landscape? Short answer, no. CWD has been around for a long time, albeit in low frequency. Similar diseases are in other animals, and it's likely to have been a condition for millions of years. In fact, it is thought that the disease spontaneously arises in about 1 in a million animals. If this is the case, why haven't we observed other species suffering ill effects from a TSE? It is possible that keeping captive deer in close proximity enhances chances of spread.

CWD came to our attention in the 1960s when sheep and mule deer were penned in close proximity for an experiment on responses to food stress. Since CWD escaped the captive environment, it has been increasing in the wild. CWD is now in at least 36 states. In Wisconsin, somewhat of an epicenter for CWD outbreak, the frequency of CWD increased from 3% in 2002 to 52% in 2017. Thus, while we were actively and consistently moni-

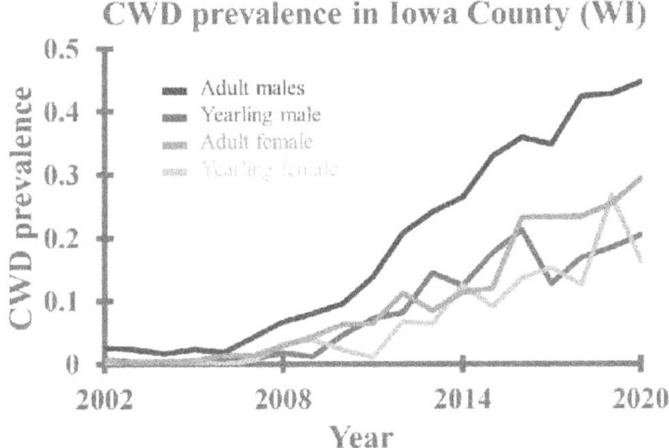

CWD prevalence by age and gender of tested cervids in Iowa county, Wisconsin over time. Data from Wisconsin Department of Natural Resources (DNR). C. Silva, 2022. If CWD was not increasing, then the number should not go up. That is, while "they" were looking for it, it went up by 10X, it wasn't that "they" just started looking or simply looked harder.

toring for CWD, it increased greatly in a decade and a half. The same is true lots of places, so it is not, as some claim, recently been found because we just started looking for it or started looking harder. For example, below are data from Iowa County, WI, showing that it is the prevalence not just the amount of testing that has increased. However, perhaps good news, in some places in Colorado, the frequency seems to be stabilized at a relatively low number (< 15%).

What to do? One of the management strategies is culling deer where CWD+ individuals are found. Today, most CWD experts think that culling will work in the very early stages of a CWD outbreak, as it will reduce the deposition of prions into the environment and reduce deer-deer contact. Some CWD experts suggest that as the disease takes hold, it is actually the accumulation of prions in the environment that poses a significant threat to deer irrespective of how many there are, and once established, culling is not the answer because the spread does not depend on herd density. Also, feeding bans are widespread, aimed at reducing what is thought to be a serious avenue for spread of infectious prions. That is, like flu and COVID-19, individuals begin shedding prior to onset of obvious symptoms.

Some conspiratory theorists think that government agencies want to keep CWD a hot topic because it helps employ more people. Seriously? Drive around a DNR parking lot sometime and look at the cars and trucks — they're not Mercedes and BMWs. Given all that the state DNAs have to manage for fishing, hunting and trapping, can anyone believe that they are hyping CWD for a few jobs? Most of them are deer hunters too. Thinking that someone without training in the science of wildlife and particularly disease management somehow knows more than disease biologists is counter-productive to the health of deer herds.

What does the future hold? CWD control is a scientific exercise, and it's a very difficult one. Even the smartest disease biologists are unsure how to control outbreaks. Part of the problem is the existence of prions in the environment, so unlike other deer diseases like Epizootic Hemorrhagic Disease or blue tongue, CWD will not "burn out" until prions are gone not just from living deer but from the environment.

18. Is It Possible That There Is Misinformation About CWD On Social Media?

Cabin Talk.—Sometimes celebrities get away with spouting things they have no understanding of. That is, they are in what I call the Dunning-Kruger Zone. This idea put forward by Dunning and Kruger states that people often don't understand enough about their topic to realize that they don't know what they're talking about. For example, more than 50% of Americans believe they are above-average drivers. A well-known NBA player apparently believes in the flat-earth myth. I mean, seriously, they have a large round spinning object in their hands hundreds of times a day. Whilst most of us just shake our heads at the lack of knowledge, others might believe such myths because they come from a celebrity – maybe that's what they talk about in the huddle?

I watched a YouTube video in which a well-known rock and roll musician and hunter explained, in his opinion, the origin and threat posed by CWD. He spoke with an air of authority and said things like "I'm aware of all this stuff but believe me when I tell you." Red flags should go up, bells should ring, because he has no degree or practical experience in wildlife management. He is, simply put, a voice lacking credibility. Because he is frequently in the public eye, and many people are apparently inclined to take him at his word, I, as should you, decided to fact-check his claims about CWD against what deer and disease scientists think. Not everything he says is wrong, much of it is.

Rock star: "CWD came from Colorado Department of Wildlife, in 1967 they injected scrapie, which is the sheep version of the spongiform encephalectomy into the mule deer in 1967 and they got out, that's how it got started."

Fact check: Unfortunately, 0% correct here. In the mid-1960s, experiments by Gene Schoonveld, then a M.S. student and now retired from the Colorado Division of Wildlife, had an experiment in which starved deer and sheep were co-penned. Schoonveld thinks that his some of his deer died of CWD, or something like it, which, in retrospect, could have resulted from scrapie crossing the species barrier from sheep into deer under those conditions. There were no injections into mule deer! Also, I, as should you, expect experts on a subject to have a handle on the language – "encephalectomy" means surgical removal of the brain. CWD is an encephalopathy, and the version in sheep is termed scrapie, not scrapie. Ok, the last one is nit-picking.

Personally, I wonder whether we might have dismissed the relationship between sheep farms and CWD too quickly. I once plotted the distribution of sheep farms and counties with CWD in Wisconsin and found a suspiciously positive relationship. So, either deer like to live in the same areas that are also good for sheep farms and the relationship is coincidental, or sheep might actually still contribute to CWD, as they might have in the original Colorado outbreak. Consider CWD is killing 19% of the mule deer in Wyoming annually. An infected deer has a 32% chance of living a year, an uninfected deer 76%.

Rock star: "The CWD hysteria is a scam. More deer are killed in Michigan every year by feral dogs than all the deer ever worldwide by CWD. There is no evidence whatsoever that CWD has ever compromised a cervid herd... Mention was made of an 18 yr old doe with CWD still having fawns, meaning to him that CWD is not always fatal."

Fact check: CWD progresses in a deer and kills it, and there is zero evidence, none, that a deer with CWD recovers. Nor do we think that a deer can have CWD but survive a normal life with a sub-lethal level. What about the 18 yr old doe that had CWD and was still having fawns? . Critical thinkers will recognize that the important question is when she might have gotten CWD, and it's obvious that it was recently; she hasn't been positive for very long. So, that claim falls in the myth category.

There are no scientists who have studied CWD who think it's a scam. There are no data I could find on the number of deer killed by feral dogs (actually just dogs). He wants us to believe that a total of only 58 deer are known to have died from CWD in Michigan, obviously far fewer than were hit by vehicles. However, we now realize that many deer are killed by hunters and predators before they showed symptoms. The truth is that we have absolutely no idea how many deer died from CWD in the wild in Michigan or elsewhere. As they say, the absence of evidence is not evidence of absence.

Rock star: "I think that there is no evidence to suggest that CWD is spread to wild populations from animals that get out of high fence farm operations where they all feed from the same trough."

Fact check: This is actually pretty low hanging fruit. Just about everyone thinks that the circumstantial evidence is too strong to doubt, namely that CWD in wild deer is usually first noticed near deer farms. Too many to be coincidental (note, I support deer farms if they are secure and animals are tested). Of course it's hard to throw a dart at a map and not be near a cervid farm. Could this celebrity be guilty of conflicts of interest? Rock star has personal ranches where captive cervids are kept and hunted, such as one "Managed for optimum health and indigenous bio-diversity, the ranch is home to world-class trophy whitetail deer, American buffalo and various exotics." For $10,000 (plus license) you can get in on a hunt for animals like aoudad, blackbuck or just plain white-tails. Pretty obvious that someone who owns and hunts trophy deer in an enclosure would think CWD in captive herds wasn't an issue. Furthermore, none of the exotics is "indigenous," and it's "bison" not "buffalo."

In another video, rock star claims there is no evidence that putting out a mineral block could spread CWD. One of the highest ranking scientists studying CWD, Michael Samuel (and his group from UW Madison) published a paper showing that prions are transferred from infected deer to mineral blocks and has recommended against using them. This could be one reason why cervid farms incubate high levels of CWD.

<u>Rock star</u>: "All exhaustive studies have concluded that CWD cannot cross between species…a person can't get it. They had "Macaw monkeys" get it but that's because they injected massive quantities into the brain."

<u>Fact-check</u>: It is true that there are no documented cases of CWD passing the species barrier and resulting in disease in humans. The Canadian study of CWD transmission to Macaque (not Macaw, which is a parrot) monkeys was important because it suggested that CWD could be transmitted to a primate, albeit not a higher primate like us. However, the study's conclusions were based on stuffing the stomachs of sedated monkeys with a huge amount of CWD-contaminated meat, because the question was

whether eating venison could result in CWD crossing the species barrier even under unnatural circumstances. There was also a brain inoculation but that was secondary. A follow up study by the National Institutes of Health challenged the findings.

The moral is to place your faith in people with credentials in the field to which they speak, in the same way you don't ask the kid changing your oil for advice on tumor therapy. That in no way implies they are always correct – the hallmark of science is that we change our minds in the face of new and better evidence.

19. What Might Happen If Every Deer Has Chronic Wasting Disease?

Cabin Talk.—One scenario wildlife managers try to avoid is triage. The advent of Chronic Wasting Disease in deer in the mid-1960s in Colorado and its subsequent spread to 36 states as of March 2025 has been met with varied approaches to its detection and control. In some areas, spread is reasonably low, whereas in others it's high. The fact that CWD spread from Colorado to 36 states should be warning enough: CWD is coming to an area near you. Many state agencies have adopted a wait and see attitude, or triage. Some states have advocated culling and instituted feeding bans. While the outcome is unknown, it occurred to me that someone ought to do a population model in which we ask the question: What might happen when every deer has Chronic Wasting Disease?

The scientific consensus is that the cause of CWD is a prion, a protein that otherwise serves a normal function, that for some reason becomes misshapen (see essay on what exactly is a prion). Misshapen proteins beget more of their kind. An infected deer will experience an ever-expanding pool of prions in its body followed by inevitable death. Depending on their genetic composition at the "prion gene," once infected the disease will progress to the terminal stage in from 1 to 3 years. If you're a deer with CWD, you're among the walking dead.

How could all deer acquire CWD? In review, fawns can get CWD before and after birth from their mothers. Infectious prions occur in the semen of white-tailed deer. Prions could be obtained by ingesting saliva from infected (late-stage) deer, perhaps on contaminated food sources. When crows and coyotes scavenge the carcass of a deer that died from CWD, their droppings contain infectious prions. Prions from decomposed deer, saliva, urine or feces, leach into soil and can be incorporated into plants and subsequently be eaten by deer. Prions remain infectious in the soil for years. As the environmental reservoir of toxic prions continues to grow, deer can be infected irrespective of whether there are many or few deer in the immediate area.

Unlike sheep, which have a genetic variant that prevents scrapie (the CWD equivalent), deer have a genotype that slows, but does not halt, the progression of CWD. Perhaps the question really is, why don't all deer have CWD?

If every deer is infected, what might happen to the population? The answer depends partly on the proportion of deer with the gene that delays onset of clinical symptoms. To simplify, deer with gene A live for 2 yrs post infection, gene B live for 2-3 years, and gene C up to 4 years. In a small sample from the Minnesota metro zone, my lab found that 70% of deer possess gene A and 30% gene B. If there is anything fortunate about CWD, it acts later in life, like cancer in humans, and death might not occur until after at least one breeding attempt. That suggests that more numerous 1.5 yr olds might be the main contributors to the gene pool, with a smaller contribution from 2.5 yr olds. Thus, the deer herd will probably consist of mostly young deer. If deer with gene B leave more offspring, the frequency of B should increase and result in a greater contribution from 2.5 – 3.5 yr olds. Gene C is in low frequency in most places (e.g., 2-3% in Nebraska). Nevertheless, if every deer has CWD, the age structure will be very different from what it is now. I should point out that some population models predict almost complete eradication of deer.

In species like big-horned sheep, breeding shifts to younger males when the old (trophy) males are culled. If deer don't live long enough to grow large antlers, our trophy rooms might be museums of buck heads past, because with shortened lifespans and breeding shifted earlier in life, very few trophy class animals will exist. In fact, were this to come to pass, if you saw a really large-antlered buck, you might let him walk because he might have a rare genotype (C) that warded off CWD for a relatively long time. Some deer naturally have relatively small antlers, like Coues deer, so maybe it won't be that bad. In my case, I've never been driven by the antlers-on-the-wall, instead I'm driven by my recipes for venison, none of which include antlers.

20. What, Exactly, Is A Prion?

Cabin Talk.—The cause of most diseases is understood. Our understanding of Transmissible Spongiform Encephalopathies (TSE), which cause mad cow disease and kuru in humans, and CWD in cervids, are caused by naturally occurring proteins that undergo a physical conformation in shape, which causes them to become infectious. These misshapen proteins are called prions (pronounced pre-ons, not pri-ons as you'd think). I am too lazy to do this, but, if you did one of those word cluster diagrams, which makes the size of the words proportional to how often they appear, prion would be one the largest words in the diagram for CWD. Most people who have heard of CWD are familiar with the word prion, because every article about CWD mentions them. I wonder, though, how many could describe a prion. I wanted to know and did some research.

Prion is short for proteinaceous infectious particle – I don't know why it's not just PI, but maybe it might lead to confusing prion with Pass Interference, Poison Ivy, Principle Investigator, Private Investigator, Personal Information, the mathematical expression "pi," etc.

Your body contains thousands of proteins that perform different functions, and all vertebrates have one called the prion protein. Although it's found in just about everything, its normal function is not fully established. Some think it might affect long-term memory, but I forgot why (joke).

We have understood the protein for a long time. Encoded in our DNA, the blueprint of heredity, is an alphabet of just four letters, ATGC, called nucleotides. The nucleotides in your DNA is divided up into segments called "genes." The string of nucleotides in a gene is read by your cellular machinery in groups of three termed codons (e.g., ACC, GCT, TGC). Each group of three is interpreted by your cell to insert a particular amino acid into a growing chain, which becomes a functional protein. The human body uses only 21 amino acids, but it's possible to make thousands of different proteins from these.

The gene in deer that codes for the prion protein exists in several forms, as a result of mutations. It's still just one gene, but with variant forms (called alleles). Some of these mutations make a deer's prion protein less prone to misfold, or at least, they lengthen the time for the progression from uninfected to clinical levels. A deer once exposed to infectious prions will enter the gauntlet of an ever-expanding pool of prions. Depending on their genotype (genetic composition) for the prion gene, the disease will progress to the clinical stage in from 1 to 3 years. If you're a deer and you have CWD, you're among the walking dead.

The prion gene in deer consists of 771 nucleotides that are translated into a chain of 257 amino acids (i.e., 771 divided by 3 = 257). Two amino acids that affect CWD susceptibility are numbers 95 (DNA base positions 283-285) and 96 (286-288). If a CWD+ deer has the amino acid H at position 95, it takes much longer to progress to the clinical stage, and if it has an S at position 96, disease progression is also delayed. These relatively simple changes affect the rate at which misfolded prions build up in the body. Most deer currently have the common sequence

ATGGTGAAAAGCCACATAGGCAGCTGGATCCTAGTTCTCTTTGTGGCCATGTGGAGT
GACGTGGGCCTCTGCAAGAAGCGACCAAAACCTGGAGGAGGATGGAACACTGGGG
GGAGCCGATACCCGGGACAGGGAAGTCCTGGAGGCAACCGCTATCCACCTCAGGGA
GGGGGTGGCTGGGGTCAGCCCCATGGAGGTGGCTGGGGCCAACCTCATGGAGGTG
GCTGGGGTCAGCCCCATGGTGGTGGCTGGGGGCAGCCACATGGTGGTGGAGGCTG
GGGT**CAAGGT**GGTACCCACAGTCAGTGGAACAAGCCCAGTAAACCAAAAACCAACAT
GAAGCATGTGGCAGGAGCTGCTGCCGCTGGAGCAGTGGTAGGGGGCCTTGGTGGC
TACATGCTGGGAAGTGCCATGAGCAGACCTCTTATACATTTTGGCAACGACTATGAG
GACCGTTACTATCGTGAAAACATGTACCGTTACCCCAACCAAGTGTACTACAGGCCAG
TGGATCAGTATAATAACCAGAACACCTTTGTGCATGACTGTGTCAACATTACAGTCAA
GCAACACACAGTCACCACCACCACCAAGGGGGAGAACTTCACCGAAACTGACATTAA
GATGATGGAGCGAGTTGTGGAGCAAATGTGCATCACCCAGTACCAGAGAGAATCCC
AGGCTTATTACCAAAGAGGGGCAAGTGTGATCCTCTTCTCCTCCCCTCCTGTGATCCT
CCTCATCTCTTTCCTCATTTTTCTCATAGTAGGATAG

(QG) that is most susceptible to CWD. Below is a DNA sequence of a prion gene from our study of white-tails in

Nebraska with the six nucleotides thought most to affect CWD progression in bold and underlined. It's pretty amazing, actually, that this string of 771 letters provides the instructions for constructing a complicated protein.

Following is the amino acid sequence of the prion

MVKSHIGSWILVLFVAMWSDVGLCKKRPKPGGGWNTGGSRYPGQGSPGGNRYPPQGG
GGWGQPHGGGWGQPHGGGWGQPHGGGWGQPHGGGGWG**QG**GTHSQWNKPSKPK
TNMKHVAGAAAAGAVVGGLGGYMLGSAMSRPLIHFGNDYEDRYYRENMYRYPNQVY
YRPVDQYNNQNTFVHDCVNITVKQHTVTTTTKGENFTETDIKMMERVVEQMCITQYQ
RESQAYYQRGASVILFSSPPVILLISFLIFLIVG

protein derived from the DNA sequence above from a white-tailed deer that is most susceptible to CWD. The amino acid code is different from the four-letter DNA code. If you recall from biology class, the instructions for making amino acids, the building blocks, come in sets of three nucleotides. The bolded, underlined letters (QG) at positions 95 and 96 indicate the amino acids that confer some degree of resistance, and they were derived from the bolded letters in the DNA sequence (CAAGGT). A resistant deer would have HS instead of QG at positions 95 and 96. In a small sample from the Minnesota Metro Zone, there were no individuals with 95H, but 30% were 96S.

It's even more amazing that this important physiological consequence for the disease stems from just a few letter changes in the code for the gene. Other letter changes make no difference to the protein structure, others do. If you're a deer, you'll want H at position 95. Actually, all individuals have two copies of each gene, so you'd really want to be 95HH or 96SS.

How does this help us understand the prion? When you cut yourself, a string of letters doesn't pour out. Proteins are not just straight strings of amino acids. Functional proteins have a 3D structure because some amino acids along the chain have more or less attraction to others elsewhere in the chain, and these attractions cause the protein to assume a shape. No you don't see these flowing out of your cut either, but if you had a great microscope, you'd see something like that shown in the image.

The biochemical jobs that proteins do depend on their 3D structures. As the image shows, there are lots of twists and turns. The protein structure provides what are essentially docking stations, places on the protein that allow interaction with other molecules that conduct your cellular processes (e.g., why you're alive reading this). Not all mutations at the DNA level cause a change in the amino acids, but those that affect the docking stations usually do. The misfolded prion protein whose 3D structure is no longer functional causes CWD and is in essence a toxin that builds up in the body (and the environment) over time, with the time to death depending on the particular

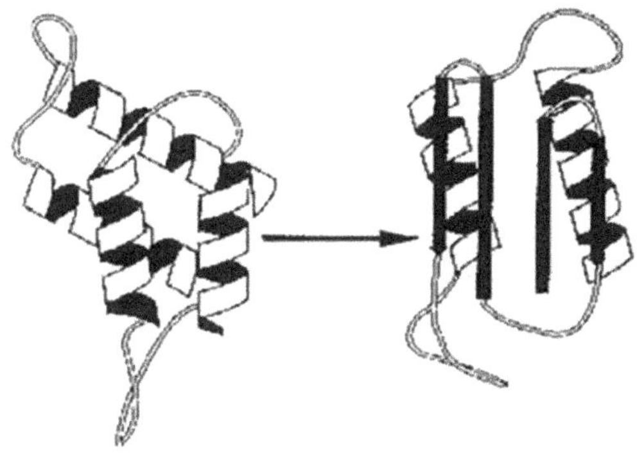

This shows what happens to a normal prion protein 3D structure (left) when it misfolds (right). Misfolded prions clog the central nervous system and are 100% fatal.
http://www.agric.gov.ab.ca/images/surveillance/pri

version of the prion gene possessed by a given deer.

To be a direct threat to us, a prion from a deer would have to be able to convert a human prion to a misshapen one, and this apparently has not occurred. However, and this make me nervous, a couple of years ago a lab took the normal prion from humans, and "challenged" it with infectious prions from deer. They were able to make the human prion convert to a misshapen configuration, although it took a lot of effort and hopefully wouldn't occur in a person who ate CWD positive venison. So, at this point in time a non-issue, but worth further review!

In theory, if deer with 95H and 96S reproduce more times than other deer, these genes should increase in frequency. Deer with the most susceptible genes might live long enough to reproduce only once. The herd will consist most likely of young individuals until the more resistant genes increase in the population.

Because prion has become almost a household word, at least among deer enthusiasts, maybe this description will demystify the genetic and physiological basis of CWD. It's pretty simple and less mysterious than you might think. Still, it's almost mind boggling that we understand the genetic basis of this disease so thoroughly.

21. Let's Expose A Bunch Of People To CWD And See What Happens

Cabin Talk.—A report out of the Canadian Food Inspection Agency found that macaque monkeys got CWD from eating infected venison. I assume given the millions of people who have eaten venison, that a large number have unknowingly consumed CWD positive meat from deer in the early stages of CWD infection, deer that appeared normal prior to harvest. The concern is whether CWD will "jump the species barrier" and infect humans, the way mad cow has. That is, will I die or get sick from eating an infected deer? Although no reports of CWD crossing the "species barrier" from deer to humans, to be sure that it won't happen, what are needed are scientific experiments.

I asked my university evolution class how you would go about scientifically testing whether CWD affected humans. A timid hand went up and the student suggested you would expose a person to CWD prions and see what happened. But, he added, it would be unethical. I responded that I didn't know of a valid scientific study in which a sample size of 1 would be acceptable. You'd need to infect, with varying concentrations of infectious prions, thousands of people of both sexes, different age groups, different genetic and socioeconomic backgrounds, among other co factors, and follow them for a couple of decades at least. This would be a difficult sell in research environments, ethics wise.

That should establish the scope of the experiment and why the proper experiments have not been done. Most experiments involve non-human model organisms, like transgenic mice. There is, however, an inadvertent human experiment underway in New York. On March 13th, 2005, venison from a CWD infected deer was unintentionally served at the annual Sportsmen's Feast in Oneida County, New York. The infected deer was tested per the New York State Department of Environmental Conservation (NYSDEC) mandate, but there was no stipulation that the test precede the banquet, as NYSDEC noted, correctly, that there is no evidence that CWD is transmissible to humans. That is, knowledge that CWD+ venison was served came after the fact. Oops.

Of the potentially 200 people that either consumed or handled the CWD positive venison, 81 people agreed to be part of a long term health monitoring program. The health of these people were monitored by RM Garruto and colleagues from the State University of New York at Binghamton, who had two objectives: 1) determine if exposed individuals had altered their "general risk behavior" and 2) determine if exposed individuals had any symptoms of prion like disease. The 81 people were sent questionnaires annually and then bi annually, unless they had died (presumably from a natural cause and not CWD!). They were asked about whether they had had contact with deer, consumed venison, killed deer, field dressed deer, butchered, and whether they wore gloves.

Questions about health included: loss of hearing, loss of vision, heart disease, memory loss, type 2 diabetes, Creutzfeldt Jakob Disease (CJD), cancer, Parkinson's disease (PD), weight loss/ gain (>5 pounds), Multiple Sclerosis (MS), hypertension, Alzheimer's disease, and arthritis. I was nervous reading this list as I have many of them, including, I think, memory loss, but I can't remember....

The news is good, so far. A 15-year longitudinal follow-up of participants up to 2020 found that venison consumption had decreased, and there were no definitive health outcomes that indicated transmission of a neurological disease to humans. There were no physician diagnosed instances of CJD, Parkinson's, Multiple Sclerosis, or Alzheimer's Disease among the final 2020 cohort (n=29), or as far as they were aware, among any of the original participants from 2006 who attended the game dinner and who were lost to follow-ups.

Although this was an "experiment" it was lacking in many ways. At the risk of being crass, this important, albeit far too small scale experiment, was messed up because "Consumption of tissues that represent a high risk of prion transmission, including brain and spinal tissue, is not known to have occurred at the feast." Seriously, if we were experimenting with mice, that would be a vital component of the study!

It can take a long time for symptoms of these neurodegenerative diseases to develop. For example, a rare prion disease called kuru is spread through ritual cannibalism among the Fore people of Papua New Guinea, but the incubation period is from two years to over five decades. It's good in that area not to annoy your neighbors. The incubation of Creutzfeldt Jakob Disease (CJD), a degenerative prion disease of people that is contracted from eating beef infected with bovine spongiform encephalopathy, is estimated to between 12 and 23 years. So the point is that continued monitoring is critical. Especially because some areas have only recently been invaded by CWD, and there has been perhaps too little time for enough susceptible people to ingest CWD prions for any consequences to become expressed.

Prions are found also in skeletal muscle (the part we eat), as well as blood, saliva and tissues of the central nervous system (brain). This suggests that at least during butchering, wearing gloves is a good idea, and it's cheap and easy. One might ask if anyone would eat a CWD+ deer? In one of my group's scientific papers on the genetics of CWD in white tailed deer and mule deer from Nebraska, 203 of 1807 (11.2%) deer were CWD+. These deer were sampled from check stations, meaning that the hunter had field dressed and brought the carcass in for testing. It is virtually certain that each of these apparently healthy deer was headed for the dinner table because no one would handle a deer with the obvious clinical signs of CWD that way. You might euthanize such an individual to save it from a slow death, but no one is going to bring it home to eat. For me, if the deer appears healthy, I assume that if it is CWD+, it is early enough in the progression of the disease so as not to be a threat. But that's my own thought unimpeded by actual data, and everyone needs to make their own decisions.

We've done, unintentionally, an experiment on CWD and people. It's not very extensive, as it should involve thousands of people from different genetic backgrounds, not just 81. So far, there appear to be no ill effects from eating CWD positive venison, which could be because it's too recent, or that it's not a threat. I think that if it were, we'd have many more cases of prion disease in people who habitually eat venison. In Colorado where CWD has been in elk, mule deer and white-tailed deer since the mid-1960s, there are no reported cases of a TSE disease that emanated from an animal with CWD. Thus, hunters are keeping their collective fingers crossed.

22. Ticks And CWD

Cabin Talk.—Few things are as irritating as waking up in the middle of the night with that nagging feeling that there's a tick crawling around on your arm or head. You reach around and feel for it, not finding it and start to go back to sleep. There, you feel it again, and once again are fully awake searching for a tick. Sometimes this repeats itself, because at least to me, ticks can be very clever when they think (ok that's giving a tick too much credit) you're after it. Then, ah ha! Success, you caught the little bugger. Holding it between your thumb and finger, you now have to admit you'll have to get out of bed and dispose of it. A few I've managed to drop in our bed when falling back to sleep. Sometimes I get my nail clippers and euthanize it. Whoever said that all living things were created in perfection is almost right, but there has to be a subcategory *but extremely annoying*. But what do ticks have to do with CWD?

I have explored elsewhere the many ways that infectious CWD prions can be transferred among deer. A relatively newly identified vector, perhaps unsurprisingly, is the deer tick. We have become aware of diseases from deer ticks because of their ability to vector Lyme Disease. However, a new study in Nature's Scientific Reports, one of the highest ranking scientific journals, found that "Ticks harbor and excrete chronic wasting disease prions." The team was led by H. N. Inzalaco from the Wisconsin Cooperative Wildlife Research Unit, University of Wisconsin.

Their work consisted of an experimental and observational phase. The first phase, although serious and important, is kind of humorous because it involved feeding trials with ticks - one wonders if ticks have a Pavlovian response, or if they all scurry to the trough when it's feeding time. They fed black legged ticks (deer ticks, or *Ixodes scapularis*) deer blood that was spiked with CWD prions. They found that the ticks so exposed ingested, and could pass sufficient quantities, of infectious prions. In case you're wondering whether ticks they used in their study already had some prions, they bought "pathogen free" ticks form the Oklahoma State Tick Rearing Facility, something probable few knew existed (me for example), prior to beginning their experiments.

In their observational phase, they found that ticks feeding on wild, CWD positive deer also ingested prions. The "wild" ticks were obtained from hunter harvested white tails in Wisconsin, which were manually examined for ticks, while wearing gloves, and any ticks were removed and bagged (separately for each deer to prevent cross contamination). The ticks were then frozen, which I imagine is one of the more satisfying parts of the work.

One of the ways we think CWD prions might be transferred from deer to deer is by mutually grooming, a common social behavior in cervids. This mutual grooming could spread prion laced saliva from deer to deer. Presumably, a deer that was groomed by another deer might also groom itself in the same spot, completing the transfer. Inzalaco and colleagues suggested a twist on this idea, in that by mutually grooming one another, deer will ingest engorged ticks on occasion, some of which might contain infectious prions that could be transferred. This could be a way in which the relatively rapid spread of CWD occurs in nature. Inzalaco and colleagues also noted that it was unknown as to whether deer stop grooming a deer that is expressing the clinical symptoms of CWD. That is, does a healthy deer avoid grooming an obviously infected deer, if so it might lessen the chances of this route of spread.

Lastly, it has been proposed that moxidectin, a modern derivative of ivermectin (think, something not to take for treatment of COVID-19), has been touted as a way to treat supplemental deer food, as it will ward off deer ticks that might be carrying CWD. One goal is to reduce the density of deer ticks and their potential to spread Lyme Disease. From a wildlife perspective it is ironic that we would bait deer with treated corn to make them tick free when baiting is viewed as a way in which deer spread CWD.

BIRDS

23. Duck Poop And Fish, Sometimes A Happy Mix

Cabin Talk.—We tend to look out our cabin windows and take for granted the plants and animals that we see. We seldom stop and wonder whether the same set of species has always been here, realizing of course that we're not talking about dinosaur age things. In recent years we have been even more aware of the fact that some animals are part of our environment because we put them there, like ring necked pheasants, which we like, or aquatic invasive species like purple loosestrife, which we don't like. Apart from what people have done to the landscape, what are the natural ways in which biodiversity shifts across the worldwide landscape? If you lived, without impact to your surroundings, in one spot 500,000 years ago, how would your environment change?

Biologists for centuries have marveled at oddities in the distribution of plants and animals. For example, moles are found throughout eastern U.S. and Canada, nowhere in the Rockies or Sierra Nevada, but occur in California, Oregon and Washington. The gopher turtle occurs along the Florida coast, coastal eastern Texas, and Arizona, and nowhere in between. Closely related salamanders, who shun salt water, are found in North America and Europe.

The red-shouldered hawk is found commonly in eastern North America but not in the Great Plains or the mountains of the west but occurs in California and Oregon and northern Baja California. Many species of wrens are found in North America and South America, but only one species, the winter wren, occurs in both North America and Eurasia.

It's not just animals. Crowberries are found in Canada and southern South America. Over 30 types of ferns are found in both Eastern Asia and North America. A massive summary of hundreds of seed plants by Robert Thorne in 1972 documented isolated populations of the same species between Africa Asia, Asian Pacific, North America, Pacific Islands, to name but a few. For example, a dwarf willow is found in eastern arctic Canada and Scandinavia. The list of species whose populations are isolated from one another by vast distances could go on for hundreds of pages. If you're wondering, it is not thought that humans moved these creatures across these vast areas distances, somehow, nature did it herself.

How is it that the same species can be found in two different places isolated by inhospitable habitat? A moment's reflection suggests some ways this could occur, and some ways that might not have occurred to you. If you put up a net high in the sky over oceans, you find small insects and spiders that are adrift in the atmosphere (actually called "insect drift"). Some animals might hitch a ride on drifting logs. Animals like birds disperse across areas not to their liking, and if they find somewhere suitable, they settle and expand. In some cases the results are remarkable. Instances in which animals or plants move across unfriendly terrain are termed dispersal. For example, the birds of the Hawaiian Islands are thought to have arisen when a flock of house finch-like relatives got blown off course and ended up in the islands, where they evolved into many new, very different looking species. The native birds of Hawaii, both living and extinct far surpass anything the Galapagos Islands, with Darwin's finches, can offer. They are not, however, gaudier than the Birds of Paradise, the champions of island splendor.

We tend to forget that areas did not always look like they do today. For example, as the ice from the last glacial period melted, the Great Basin, which is today arid and hot, was a swamp. At the last glacial maximum 21,000 years ago, sea level was 400 feet lower than today, which exposed land that was previously under water. This meant that "land bridges" appeared that were formerly submerged and provided highways for animals that don't fly to spread out to new areas. Subsequent warming and melting of glaciers resulted in the rising sea levels and covered the bridges and leaving land dwelling species in two different places. In these cases, the different areas of residence did not result from dispersal, but rather from the disappearance of suitable land in between ends of a once continuous distribution (a process known as vicariance if you wanted to know).

I am constantly guilty of "burying the lede" and have done so here. What does dispersal have to do with species that live in water? Fish are often found in remote lakes or ponds. Maybe two lakes are close together, but to a fish, ten yards of dirt is like an ocean to a shrew. Why do so many lakes in Minnesota have walleyes (apparently excluding the ones in which I fish)? Everyone knows that rivers and streams provide aquatic highways that connect lakes. However, we also need to consider the past. As the last glaciers melted, lakes and rivers were much more extensive than they are today. Glacial lake Agassiz covered almost 116,000 square miles in Manitoba, Ontario, Saskatchewan, Minnesota and North Dakota, whereas Lake of the Woods today covers 1,679 mi^2. As glacial lakes drained, they left smaller lakes behind separated by of exposed land and isolated fish populations, like walleyes. This is "vicariance" in the aquatic realm.

Animals transport other animals as well as plants. Birds like American Robins eat mulberry seeds, move off, and the seeds can be viable in their droppings. Ducks transport influenza virus like H5N1. Vultures can poop out viable anthrax bacteria (and it doesn't kill them as it passes through their digestive tract). Note to self, avoid areas around vulture nests.

One of the more bizarre and unanticipated examples of birds transporting other species across distance was published in the prestigious Proceedings of the National Academy of Sciences by Adam Lovas Kiss and colleagues from Hungary and Spain. They knew that eggs of the killifish, an egg-laying toothed carp related to the live bearing guppy and swordtail, could pass through swan guts and remain viable. What about other fish eggs?

Most assumed that it was obvious that once in the gut, eggs were digested and by the time they were passed out of the GI tract, the fish eggs were goners. But, why let conventional wisdom stop you? Lovas Kiss and his group force fed captive mallards eggs with living embryos from common carp and Prussian carp. They wanted to know if the embryos would survive passage through the mallard gut and come out the other end alive. How would they know?

You can't just follow mallards around with a specimen cup and hope to catch duck poop, as glamorous as that might sound. They kept the ducks in cages with plastic trays underneath, and 1, 2, 4, 6, 8, 12 and 24 hrs after force feeding the living fish embryos to the ducks, collected the mallard droppings. They immediately put these in local river water where the fish had been obtained.

Contrary to common sense prediction, they found living embryos after being pooped out of the duck! But, I hope you weren't expecting a huge success rate. Only 8 common carp eggs (0.2% of those fed to the ducks) and 10 Prussian carp eggs (0.25% of those fed) survived. Some of the embryos later died from fungal infection, despite their efforts, but a couple of each carp species actually hatched.

It might be hard to be impressed by this sort of success rate. However, imagine thousands of migrating mallards, especially in the bygone days of duck numbers. I can easily imagine that many fish would be transported to new locations, although most of the embryos were recovered in the first hour after being ingested, which means they couldn't have been transported far even by a migrating duck. Other studies have found as many as 217 fish

Drake Mallard. Image: Roland Mey

eggs in a single mallard, and over 63,000 in a single gull.

Since this study appeared, two others came out that describe equally bizarre gut passages. S. Sugiura, an ecologist at Kobe University (Japan), found that when a frog eats a water beetle, the beetle swims through the gut, and climbs out the frog's rear end. Wow, those beetles are off my diet if stranded on an island. In another study just released, Silva and colleagues found that floating plants called Wolffia, the smallest of all flowering plants, can pass intact through the guts of swans and ducks. The same authors found experimentally that "killifish eggs from bird feces were capable of continuing their development after spending over 30 h inside a swan".

These observations reveal how it is possible to explain why some species of fish are found in unconnected (today) bodies of water – they got there via duck or swan poop. All of this adds to our understanding of how plants and animals come to have some bizarre distributions. The frog story is particularly bizarre, because most of us thought that once your meal was inside your gut, it was yours. Like looking in your toilet and seeing a snake head, only in reverse.

24. What's In A Feather, According To The NEW Law?

Cabin Talk.—A lot of arcane laws exist at state and federal levels. How do I know this? I googled it. I learned that in Alabama, it's illegal to wear a fake mustache in church that causes "unseemly" laughter. Nebraska makes it possible to arrest parents of a child who burps during church services. In Rhode Island, you cannot throw pickle juice at a trolley. Besides arcane laws, what about those that are designed to protect our wildlife but seem over the top? You innocently find a feather on the ground and stick it in your hat, is that ok? There have been some substantial changes in late 2024 to how the Migratory Bird Treaty Act is enforced.

The Migratory Bird Treaty Act (MBTA) says that it is illegal for you to have in your possession a bird, part of a bird, nest, egg, of anything except a starling, pigeon, or house sparrow, or game bird you harvested during an open season (might need to show you had a hunting license). But the new regulations provide a caveat.

Previously, it was illegal to be in possession of a bird, part of bird (feather), nest, eggs, of anything covered by the MBTA. If you find a bird nest in middle of winter, and it would make a nice wall ornament, forget it, it's illegal. However, Federal and State authorities basically looked the other way if a private citizen found a dead bird, bagged and froze it, and brought it to a museum or nature center, institutions with authorization to accept the specimen. Note, that means that you cannot put the nest you found in winter on your wall. These institutions obtained many valuable research and teaching specimens; hence, it was good. But the person finding the bird was still technically in violation of the MBTA by having the bird in his/her possession. If someone wanted to do this routinely, they needed to be added to a permit form a museum or get their own permit, navigating the federal permitting gauntlet. For example, my colleague and I sponsored a program in downtown Minneapolis and St. Paul, where for 10 years volunteers walked a prescribed route in spring and fall migration and picked up dead birds, recoding where and when they found them. Each volunteer had a letter from me authorizing them to salvage under my scientific salvage permits. The goal was to document what birds were killed and which sorts of buildings were the biggest threats to migrating birds.

For the general public, people would say when they were in possession of a bird part, that I just found the feather. Didn't matter, as the wording of the MBTA says that it's illegal to have the feather or bird part in your possession. It was not up to a federal or state agent to prove how you obtained the bird part, simply possessing it was a violation.

In late 2024, the USFWS rewrote how they would handle enforcement of the MBTA by ruling that anyone without a permit could salvage migratory birds and their parts. This removes the look-the-other-way that law enforcement had to use. However, and this is huge, it is now a requirement that if you find a dead bird, and want to donate it for research or teaching, you can only possess the bird or its parts for 7 days, otherwise you are in violation of the law. Seven days might seem long, but it's not to most, as it usually requires finding a repository and making an appointment to drop it off. Thirty days would have made sense.

Here are some other aspects of this issue:

Q. Can I salvage migratory birds for personal use?

A. No one is allowed to salvage or possess migratory birds for personal use. All migratory birds salvaged must be transferred to a public scientific or educational institution, zoological park, museum or scientific society as defined in 50 CFR 10 or a Migratory Bird Special Purpose Possession permit issued under 50 CFR 21.27.

Q. If I have a Federal salvage permit, do I need a State permit to salvage migratory birds?

A. Your Federal salvage permit is not valid unless you are also in compliance with State requirements. This means that if your State requires you to have a permit to salvage birds, you must hold a valid State permit in order for your Federal permit to be valid.

Q. Do I need to tag the birds I salvage?

A. Yes. Each migratory bird salvaged must be tagged with (a) Date and location the specimen was salvaged, and (b) Name of person who salvaged the specimen. Other details like likely cause of death are useful, e.g., hit a window, killed by vehicle.

The most common response I hear is this: "I found this red tailed hawk feather and stuck it in my hat. Surely that's not illegal, I didn't kill the bird, I found a molted feather. You'd have to prove I killed the hawk and you couldn't because I didn't." Sorry, the law says it's illegal to possess and retain this feather, past 7 days, without a permit. The burden of proof is not with law enforcement to prove how you obtained it, possession equals violation.

Sounds a bit over the top? Think about wearing your hat with the hawk feather sticking out of the band and being confronted by a game warden. The warden knows that someone in this area shot several hawks illegally. You say "Hey, I just found this feather on the ground

Townsend's Solitaire specimen found in downtown Minneapolis stored in ornithology collection at the Bell Museum, University of Minnesota.

and stuck it in my had." The warden has no way of knowing whether it was you who did the killing and if the feather in your hat (pun not intended) was from those dead birds, but the law makes it so that they don't have to guess. Possessing the feather is a violation of the law irrespective of how you obtained it. Now, they will probably let you off with a warning, but it would be within their jurisdiction to charge you with violating the MBTA.

The law was enacted in 1918 because people were killing large numbers of herons and egrets at breeding colonies for their plumes to be used in the millinery trade (see photo). But the law protects against more than illegal

Photo National Audubon Society

feather harvest. When I was in the Museum of Natural Science at LSU in Baton Rouge, a game officer came to the museum and wanted to know if the feathers he had were from "gros bec." I didn't want to appear stupid, I was a museum curator after all but being from Minnesota I had no idea what a gros bec was. I assumed it was a Cajun name for a Rose breasted Grosbeak, but it was obvious the feathers were not from that species. After sifting through the bird skin collection, I matched the feathers to a yellow crowned night heron, which have "big beaks," and hence, gros bec in Cajun French. The warden was part of a sting operation that caught a number of people illegally killing the night herons at a breeding colony and selling them for food (going rate I was told was $2-$3 per bird). Apparently gros bec was a local Cajun delicacy (the birds eat a lot of crawfish).

We might think, then, since these egret killing sprees are over, we can roll back the law. But, if were legal to kill birds, I firmly believe that we would see the rebirth of target practice on birds by hunters at least in the early season. In fact, Hawk Ridge in Duluth, Minnesota, where thousands of hawks migrate past in the fall, was a popular spot for hunters to go and shoot passing hawks to warm up for the waterfowl and grouse season. The "few bad apples" theory would be the reason for keeping the law. Everyone with a driver's license reading this has been guilty of speeding, but most of us are ticketed rarely only when we really speed. Same with the MBTA.

OK, what *should* you do if you find a dead bird in decent condition? It comes down to what is meant by "possess." If you hold on to it and keep it for yourself, you are in violation, irrespective of how you obtained it (e.g., even if you found it). If you put a dead bird you found in a good plastic bag with a label, freeze it, and transport it to a repository with salvage permits (e.g., museums, some nature centers) within 7 days, the Feds or State are not going to come after you, they have better things to do.

In all the museums in which I've been a graduate student or curator, over four decades, many specimens are brought in by the public. USFWS and state agencies are ok with this, as the specimens don't go to waste and can be scientifically valuable. One of the few specimens, if not the only, of Townsend's Solitaire in Minnesota was picked up in downtown Minneapolis and brought to the Bell Museum.

What should you do if you find a really stinky, rotten carcass crawling with maggots? Probably just leave it, no one else wants it either. However, if it's really valuable, even a foot or a few feathers can be important, as we can extract DNA from a foot or feather. And don't fret about leaving common things lay. Nature has evolved an impressive set of scavengers, from mammals to maggots.

One last caveat, if you find an eagle feather then call your local DNR because even permitted museums cannot accept them. A single feather can be a good educational opportunity: what species is it from, on what part of the bird did it come from, was it an old feather and molted or from a predator kill?, sex?, age?. If you want to know what kind of bird it came from, use the USFWS feather identification website (https://www.fws.gov/lab/featheratlas/idtool.php). I hope this helps clarify the MBTA and the public a bit more.

25. Do Bird Feeders Help Wintering Birds?

Cabin Talk.—Deep dark winter days in the Northwoods pose survival issues for many animals. Looking out my cabin window on a snowy day makes me realize I'm glad I'm inside – yes, I could bundle up and survive if I found shelter and maybe some firewood, but I'd miss the game that's about to start. I always find it almost comical that bird watchers come inside when it's too rainy or cold, and I wonder what they think the birds do, just call it a day and come inside too? How much should we help our native wildlife, especially during winter, when the weather exceeds our comfort zone?

Many bird species use feeders during winter. Besides providing some interesting times watching their interactions, does it help or hurt the birds? Within a couple of square feet at my feeder trays there will be several individuals including American goldfinch, house sparrow, purple finch, dark eyed junco, northern cardinal. These crowded densities, hour after hour, throughout the winter,

A blue jay about to depart after eating 27 pieces of cracked corn, and a female red bellied woodpecker dining on suet.

occur rarely, if ever, in nature. Because of the typically extreme density of food at the trays, aggression levels are relatively low, by no means absent, but because everyone can eat their fill, why fight for access to a superabundant resource? That is, the normal territorial interactions that keep birds separated for most of the year are abandoned because it's not economical to be territorial when food is abundant.

The concentrations of many species that come to feeders within inches of each other are not natural. Obviously, gulls, ducks, etc., regularly hang out in dense aggregations, almost neck and neck, during on breeding periods. Even during breeding, their nests are just a pecking distance apart. And we sometimes observe birds that come to feeders in close proximity in nature during migration and winter, assuming high food availability. For example, flocks of tens of thousands of red-winged blackbirds are seen in migration and winter, but at my feeders, there are rarely more than three or four; any more starts some aggressive behaviors. But seeing nuthatches, juncos, chickadees, cardinals, sparrows, blue jays, and others feeding within inches of each other is pretty not normal. Suspension of agonistic interactions at feeding stations shows an adaptability in our winter birds, although not all species are friendly.

The larger species, including blue jays and the introduced European starlings, often chase off smaller bodied species. I suspect blue jays of using the occasional hawk scream call to clear the landing platform. One winter we had a pair of yellow shafted flickers that no one messed with. I have seen a European starling drive a red-bellied woodpecker off the suit cake. A blue jay swooping in from the woods often results in the tray erupting in a panic, as it's not clear what the heck it is. The feeding birds want to be sure it isn't an adult Cooper's hawk looking for a bird that has dropped its guard, so vigilance and the occasional panicked flee makes sense. The hawk gives another meaning to "bird" feeder.

And the food we feed winter birds, how natural is it? Sunflower seeds are quite popular at my feeders, but these are seeds from the domesticated sunflower. Native sunflower species ranged across much of North America, but the natives don't have huge seed heads like those in the domesticated versions. It is virtually certain that in Northwoods winter birds did not have access to copious piles of sunflower seeds throughout the winter. Nor piles of other seeds, including thistle.

Woodpeckers love suet. And they have a pecking order at my suet feeders in order of size. Everyone chases off downy woodpeckers, red-bellied and hairy woodpeckers seem to have a truce, although I've not seen many interactions. Unfortunately, here at our place in Nebraska, we have no pileated woodpeckers at the feeder, at least yet. But, is the concentration of woodpeckers "natural"? I'd say not. First of all, Nebraska was once prairie with very few trees. But not now. The question then is where in nature would you see all of these species of woodpeckers, plus starlings, house sparrows, etc., eating the exact same thing (suet) at the exact same place? And the winter-long availability of suet is far from the normal, pre bird-feeding era. Picking up seeds on a cookie tray alongside others of the same and different species hardly mimics natural conditions.

Although our feeder birds seem well adapted to eating seeds from a feeder, it's not their normal foraging behavior. White-breasted nuthatches and black-capped chickadees are not ground foragers, yet they regularly gulp up seeds on the flat tray out in the open, a far cry from their normal foraging sites. It indicates to me that these species are quite adaptable, being able to switch to foods they wouldn't otherwise have access to, eat in unusual settings, and tolerate unnatural densities. Perhaps this adaptability explains why the species around today were able to make it through the last several glacial cycles when climate and habitat were constantly changing.

The Harris' sparrow is a welcome addition to our winter bird group, and I can't say why, but it's my favorite winter species. Maybe because you have to go so far north to reach their breeding grounds (the southern limit of their breeding grounds is far northern Manitoba and Saskatchewan). These sparrows are in a group of "crowned sparrows" including the white-throated and white-crowned species. They along with juncos and fox sparrows have a peculiar behavior in which they rapidly kick their feet out behind them to move leaf litter on the ground and expose food items. This behavior is termed "double scratching" and is highly stereotyped. On the feed tray, we often see our Harris' sparrows double scratching in an inch or more of seeds, utterly unnecessary because even a blind sparrow could gorge itself. But you do what your genes tell you to do, and I guess it couldn't hurt to stay in practice. The seeds they knock onto the ground are seized upon by other species.

The feeding of birds, however, carries at least one negative potential problem. Unnatural densities facilitate the spread of disease. In the eastern U.S., feeders concentrated the newly introduced House Finch, and they developed a conjunctivitis (from poultry!) that spreads when birds were in close contact. Also, think of deer and CWD. Feeding bans are used to keep densities down to prevent spread of CWD and other deer diseases. On the other hand, interest in bird feeding encourages knowledge and conservation of wildlife, including donations to governmental and NGO conservation organizations and fuels a whole bird-feeding economy. Not to mention citizen science programs like the Christmas Bird Count (https://www.audubon.org/community-science/christmas-bird-count).

All in all, the birds at our feeders and what we feed them depart from the situation in nature. They are in close proximity in unnatural densities eating foods they most likely wouldn't have access to throughout the winter. The scientific question is whether it matters to populations. I'm of the impression that feeding birds all winter results in a larger density of birds than you'd have in areas without feeders. That is, I suspect that if you did a census of white-breasted nuthatches in areas with and without feeders, there would be a larger breeding population in the feeder areas. But my intuition is hardly evidence. In a 1992 paper in the J. Field Ornithology, Brittingham and Temple found that the survival rates of black-capped chickadees were the same in areas with feeders and without. So, even though I see more black capped chickadees at my feeders in winter than I see on the property the rest of the year, it's due to chickadees coming in from neighboring territories. Therefore, I think that feeding birds is mostly of value to us. Although there are negatives, feeding is probably more or less neutral to birds.

26. Do Grouse Eat Turkeys? Searching For A Silver Bullet

Cabin Talk.—We humans have a very high opinion of ourselves, and our ability to solve problems. In our quest to provide answers, we are drawn to simple explanations. There's actually a term for this, the principle of parsimony, or Occam's Razor. It basically says that the simplest explanation is often the best one. But how many problems are simple? How many are really either black or white? Maybe 1 in a 100? I guess it's something like this - a single idea might explain the largest amount of uncertainty about an issue, but if you want to understand the entire issue, don't stop with just one factor. And beware of the quest for silver bullets. I'd wager that few questions about nature have a simple answer.

It seems to me that the basic state of human understanding is ignorance, at least that's how we come into the world. You personally learn things by direct observation, reading, from teachers. But some things we "know" we heard from others who themselves couldn't really explain these things. Call it third-hand pseudo-knowledge. For example, we know that smoking causes cancer. But few can explain what smoke does to our cells, how cancers form and why some kinds of smoke are more dangerous

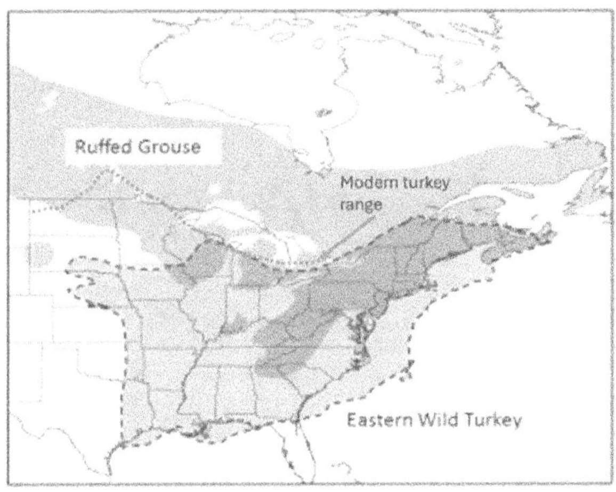

than others? Now in this case, you're on safe ground because of countless verifications of the smoking cancer link. The point is that what we "know" about this topic is a placeholder for information stored in a textbook or in some expert's head – not from our direct understanding.

But even valid information gets twisted when handed down too many times.

A frequently heard myth is that wild turkeys eat ruffed grouse eggs and are suppressing the grouse population. Turkeys are big birds, sometimes pretty feisty and aggressive, and are known to chase people around at times. They are very successful, expanding their range greatly, helped by human transplantation and releases. They seem to eat anything, so why not grouse? Someone told you this and it sounds good. Is it true or hearsay?

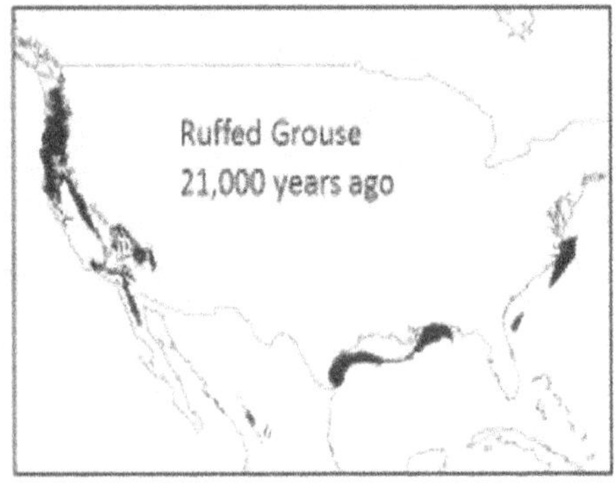

The pre-settlement range for the eastern wild turkey (see map 1) did not include northern parts of the current range (I know some disagree) and turkeys have spread because of frequent introductions (see red dotted line).

Therefore, it's fair to say that turkeys are an exotic species in several northern states like Minnesota, and we are well aware that exotic species can do considerable harm to native species. If there's a decline in ruffed grouse and an increase in wild turkeys, it's human nature to jump from correlation to causation — the "ah-ha" in the silver bullet quest.

The danger in interpreting correlations as cause and effect is well known. There's even a book highlighting many. For example, the divorce rate in Maine is almost perfectly correlated with the per capita consumption of margarine. You'd raise more than a few eyebrows if you claimed there was a meaningful connection there.

Given that we can consider turkeys as an exotic species in northern Minnesota, could they harm grouse populations? One answer is to look elsewhere where turkeys and grouse coexist. On the map we can see that eastern wild turkeys and ruffed grouse coexist in many areas outside of Minnesota. I wondered if people in areas where the two species have coexisted for a long time had the same notion about grouse eating turkeys. One such area is West Virginia, and here's what their DNR says: "Large turkey flocks have led to rumors that turkeys are eating all the foods of ruffed grouse and white tailed deer. That is not true. If an area ever has more turkeys than deer or ruffed grouse, it is because the habitat is better for turkeys than these other two species."

Another way to look at the history that the two species have had together is to use an ecological technique called niche modeling. For both of these species, I used niche modeling to estimate where they lived at the Last Glacial Maximum. Recall that 21,000 years ago, there was a mile thick glacier atop much of the Midwest, so plants and animals were displaced southward. In the estimated distributions at that time, we see that there were consider-able overlaps in the distributions of the two species where and when they rode out the last glacial period, far to the south of their current range (note the extended land areas at that time). The overall history of the two species then is one of coexistence. They have adapted to live successfully in the presence of each other.

Therefore, if turkeys killed grouse, they wouldn't coexist anywhere, grouse would be extinct, which is clearly

Ruffed Grouse. Image: Alain Audet

not the case. If competition were too intense, they would have non-overlapping distributions. Instead, they coexist and have coexisted over many thousands of years. There-fore, there are no obvious effects of turkeys on grouse populations.

Invoking the Dunning Kruger effect, we should ask someone claiming that turkeys negatively influence ruffed grouse to show us the evidence. Otherwise, the person is simply repeating what someone else said who had confused correlation with causation. We can predict that at population highs, there will be little reason to blame turkeys because, after all, there are a lot of grouse. At grouse population lows, it will be fashionable to blame turkeys. For the turkey blaming folks, how do they account for the observation that, for example, 2020 was a good grouse year in Minnesota, if turkeys are so bad?

Currently no scientific studies make the claim that turkeys are suppressing grouse populations, although one could design a study to test this. For example, one could find an area where both turkeys and grouse are fairly common. That is, an area that everyone would agree is both great grouse and great turkey habitat. Then you could put out fake grouse nests baited with grouse-sized eggs, set up trail cameras, and see if turkeys are grouse-egg bandits. My guess is that they are not, but we should do the experiments. Until then, we should stop spreading rumors without tangible evidence.

Other folks claim that wild turkeys eat bobwhite quail eggs. Dr. John Carroll, galliform researcher from University of Nebraska-Lincoln, told me that continuous video of over 800 nests in the southeast showed zero depredation by turkeys and "ironically a number depredated by deer, and almost none by coyotes."

What's with the title of this essay? Well, nothing other than trying to start a new false rumor.

27. Do Eagles Catch Wild Geese?

Cabin Talk.—Most people see the bald eagle as a majestic emblem for our country. I have no quarrel with that. If you asked these same people they would probably also say, without actually knowing, that the bald eagle is a fierce predator, and that anything from a large fish to a small child ought to beware. Truth is that bald eagles are large, majestic, fierce-looking vultures. Not taxonomically related to vultures, but ecologically very similar. The bald eagle motto is: I never met a dead fish I couldn't catch.

We often see pictures or accounts of bald eagles supposedly catching ducks, mostly on the water. Any attempt to downplay the hunting process of the bald eagle is usually met with a stream of anecdotes from people who have seen bald eagles take geese out of the air. The question is whether bald eagles are predators (not scavengers) of healthy waterfowl. I've seen quite a few bad eagles but I decided that the best source of information was going to come from professional waterfowl biologists.

I raised the issue of eagles killing geese on a list serve that is home to professional waterfowl biologists and generated many interesting responses. Nearly all of the accounts of eagles attacking geese were anecdotal. One well known waterfowl biologist commented "Three of us witnessed an adult bald eagle "fly down" an adult snow goose in South Dakota one spring. The goose was part of a large flock moving out to feed, maybe 500 or 600 ft elevation. The eagle cut the bird out from above the flock, never touching it, and shepherded it down to the ground with some masterful flight movements. Although I had no way to determine the health status of the goose, due to the confusion, it appeared as if the eagle selected the individual based on a somewhat random instantaneous availability among the flock. Geese were literally boiling around in all directions and appeared to be confused as to the exact location of the eagle. It was a very amazing predatory event to witness. Once on the ground the eagle landed on the goose which put up little struggle." And he further commented that "The eagle wouldn't risk an aerial struggle with a snow goose and wouldn't possibly risk a strike on a larger Canada goose in the air. On the ground might be a different matter." There was no comment as to whether the bird was in fact slightly injured, and if that's how the eagle chose that particular bird. This waterfowl biologist has 55 years of field experience.

Dr. Al Afton, a famous waterfowl biologist, commented to me that "I have frequently seen eagles fly low over goose flocks and scatter them, and if there is a cripple or "very slow to fly" individual, the eagle pursues them and have killed some Canada geese in my presence. Eagles do those flights over ducks and coots as well, with same results. I would note that geese in general are very skittish and take flight when small planes fly near them and I suspect that evolved from predation attempts by eagles?" Of course, geese haven't been hunted by people with guns for more than a couple of centuries, so eagle flyovers possibly evolved to flush sick or injured birds.

Another biologist wrote "It has been about 30 yr ago and I've forgotten the eagle species, but I saw an eagle take a cormorant out of the air on a WMA in western Montana." In other words, in the ensuing 30 years there were no eagle attacks on geese observed. Another biologist saw an eagle standing on top of a tundra swan but did not know how that came to be (and the eagle and swan flew away).

An owner of a waterfowl hunting club in central Nebraska said he's witnessed an eagle attack on a goose once in 40 years of hunting almost every day during waterfowl season. "A lone blue goose flew into our spread … tried to hide in the decoys and the eagle got him!" I suspect that a lone blue goose in spring migration where they have hundreds of thousands of friends was injured and not able to keep up, to its demise.

Although few waterfowl biologists had examples of eagles taking down Canada geese, there are a couple of exceptions. First, golden eagles are more predatory than bald eagles, and there were accounts of this species attacking geese, especially Cackling geese. The Cackling goose is a small version of the Canada Goose (although not that closely related). One biologist wrote to me "In NW Oregon and SW Washington it is not uncommon to watch bald eagles both adult and juvenile take healthy geese out of the middle of a flying flock of cackling geese. I have personally witnessed this on over ten occasions." Maybe the small size of cackling geese makes them vulnerable. Again, whether these were healthy geese is unknown.

Adult Bald Eagle. Image: Stacy Vitello.

One person had two accounts of golden eagle attacks, one on a field decoy, one a sitting Canada goose, over 40 years of observations. Some observers reported that eagles are predators on goose nests, goslings, and even incubating birds. One person who didn't have a waterfowl eagle observation recounted an eagle plunge into a flock of coots (as did Afton as noted above).

One biologist saw an immature bald eagle take a snow goose in flight from a flock it had scattered by flying over it and "The eagle obviously could not carry such a load, but was able to exercise a controlled decent that kept it in contact with the goose." Another wrote "A few years back I witnessed a predation event during which a bald eagle hit a flying snow goose in a very large flock. The eagle struggled to fly off with the goose and failed, so it then sat in the shallow water (flooded rice field) and began plucking and feeding on the goose. My point being that a bald eagle is likely tough enough to take down a Canada goose, so I suspect that preying on one would be more than possible. A goose having been shot would be even easier."

In general, an eagle cannot lift off from the ground with more than about 50% of its body weight in its talons. Given the years of experience, it is interesting that the author did not have any observations of this. The question really isn't does it happen, but how frequently does it happen? Is there any science on the topic?

A study of eagles killing geese was published in 1994 in the journal The Wilson Bulletin, entitled "Predator prey interactions between eagles and Cackling Canada and Ross' Geese during winter in California", by Scott McWilliams and colleagues. They wrote "Bald Eagles were observed killing Cackling Geese only in the Klamath Basin where they were responsible for 14% of all eagle kills observed". The obvious question is, 14% of what, a million or 20? Turns out the number was 21, of which three geese (14%) were killed by bald eagles and 18 by golden eagles, over three years and 155 days of observation during the hunting seasons in central California wintering areas. So the major culprits were golden eagles. The authors further commented "In 1987, we saw no bald eagles kill a cackling goose, while golden eagles preyed heavily on cackling geese." The obvious question is what does "preyed heavily" mean? Turns out a total of 12 cackling geese.

Was there any information on whether geese taken by eagles were injured previously by hunters? No, because that's ridiculously hard information to get! But the authors noted that "Geese crippled or killed by hunters may provide more susceptible prey for eagles," although an eagle eating a dead goose is not depredation, it's scavenging. Thus, of the few of geese seen taken by eagles out of tens or hundreds of thousands of geese, the authors did not know how many, if any, were wounded. However, the authors note that by the end of their observation period in fall of 1987, there were at least 18,000 Cackling geese harvested by hunters, which suggests to me a strong possibility of these geese carrying a stray shot or two.

Do waterfowl carry shot? A study published by Peter Hicklin and W. R. Barrow in 2004 noted that of 111 live Canada Geese examined with a fluoroscope from Prince Edward Island, 35% had pellets embedded in their tissues (average 2.3 pellets per bird). For comparison, 11% of 104 mallards from the same area were carrying shot (1.3 pellets per bird). Yikes, we need to shoot better. But for certain, many flying waterfowl could be at least somewhat hampered by shot, making them more inviting targets for eagles.

What I took away from these interesting accounts is this. These professional waterfowl biologists have between them over a hundred years of field observations of thousands of eagles and millions of Canada, cackling and snow geese. If bald eagles routinely hunted down healthy flying geese or ducks it would be common knowledge and there would be a plethora of accounts. Yet there are just anecdotal reports. There is no doubt that an eagle, especially a golden eagle, will attack smaller geese (cacklers) but it is fairly rare. Additionally, no one knows whether these geese were carrying shot - a goose might not have to be slowed much by embedded shot to be recognized as vulnerable by an eagle.

I'll end with a statement by one of the professional waterfowl biologists: "My opinion is that first and foremost eagles are opportunistic scavengers. Rarely in my experience do they hunt and kill."

28. A Day In Spring Migration At Leech Lake, Minnesota

Cabin Talk.—The seasons are vastly different in the Northwoods. The two biggies, summer and winter, are transitioned by spring and fall, with their own characteristic sights and sounds, especially bird migration. There are a set of species characteristic of our place in both summer and winter, with some overlap. But it's the spring migration that is perhaps the most exciting, as it signals the end of winter and the start of summer (and open water fishing). Life is returning to the Northwoods.

I reflected on a day I spent watching migrant birds along Leech Lake on a recent spring day, marveling at the changes the seasons bring. I sat along the shore on a cool late May morning watching a parade of migrant warblers passing through the partially leafed out trees. A slight breeze from my left to right along the shore was pushing emerging insects in that direction, which is the direction

the birds were mostly moving. The Canada Warbler I saw had just come from South America, using its
amazing navigation system, probably flying over the Gulf
of Mexico after molting and fattening up on the wintering
grounds, refueling in coastal Louisiana, and making its way
to claim a territory, either nearby or to the north of me.
What a feat, one I surely couldn't duplicate.

Quite a few Tennessee Warblers, rather poorly

Blackburnian Warbler, Canadian-Nature-Vision

named, moved past. So many questions. Are they related?
Are they part of the same breeding population traveling as
a cohort? Are they all males or all females? Are they first
year birds returning from their first experience in South
America, adults only, or a mix? Or are they just the com-
monest species passing by at that time, and this lakeshore
was just a good place to be migrating if you're a Tennessee
Warbler? And those plumages, which are frustratingly var-
iable to birders. Some birds have an eye stripe almost like a
Red-eyed Vireo, others tending towards being yellowish.
Some with quite bright green backs (even to me, a partially
red-green color blind male). Having looked at a few hun-

dred specimens in the hand, it's clear why they present identification challenges.

A male Blackburnian Warbler made me stop in my tracks. Spectacularly colored bird, maybe the jewel of the North Woods. Not far behind is the American Redstart. Three male yellow warblers, who stay here and raise families, were fighting continuously, each competing for the best territory. On the other hand, Song Sparrows and Chipping Sparrows had already resolved territories. And for flycatchers, I'm think one was a Least Flycatcher, another a "Traill's" (Willow or Alder). Ecologically important for sure, like the Warbling Vireo that was part of the mix, but not on a par with the other eye candy. A Caspian Tern passed just offshore and momentarily distracted my warbler watching.

One male Blackpoll Warbler flitted by. Used in my ornithology class as a champion long distance migrant, it was nice to see this bird. According to the Cornell website, they have lost almost 90% of their population over the last 40 years. Such statistics often cause much angst. But it is likely that at any one time pre-humans, some species abundances changed dramatically naturally and recall that 99% of all species that ever existed are extinct. The extent to which the decline of the Blackpoll Warbler is because of humans is unclear. Habitat loss is cited as a major factor. Also, its migratory journey, sometimes a three day, 1800 mile nonstop flight across the Atlantic, from the Maritime provinces to wintering grounds in South America presents formidable challenges. Perhaps global climate change has made the weather conditions they face on the overwater journey now even more treacherous, and without some changes to their migratory route, their fate is sealed. However, there are approximately 60,000,000 Blackpoll warblers estimated to exist, and hopefully they can recover.

Seemingly in an instant, bird activity ceased. Had all the migrants passed through? Did summer instantaneously arrive, it's the Northwoods after all? I spotted a bird a bit larger than a robin perched on a dead branch in a tall

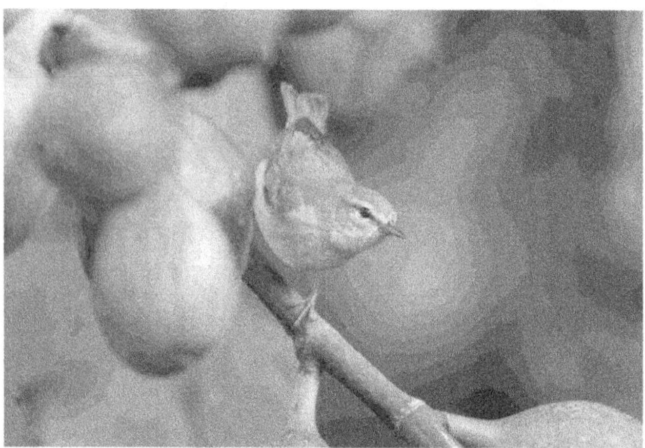
Tennessee Warbler, Francisco Javier Corado

birch along the waterfront. It was a Merlin, a bird eating falcon. Looking for breakfast. Unfortunately, predators get a bad rap because people watch too much Disney, where everything is tranquil and happy. Sure, it would be unfortunate for a Blackpoll Warbler to have migrated 1,800 miles south to winter, survive the winter (with new plants and foods, different weather, new competitors and predators), and migrate back north to end up being eaten by a Merlin a hundred miles from its intended nesting site. But it would be a win for the Merlin. Human emotions too often cloud a common sense understanding of how nature works. Raptors eat birds, and it's irrelevant whether you like it or not. Merlins and Blackpoll Warblers (and others) have co-existed for millennia and neither is extinct. You should enjoy a Merlin eating a Blackpoll Warbler as much as you do marveling at the latter's migratory feats. If not, you're not a naturalist.

Perhaps as a commentary, a Tennessee Warbler on a small branch, faced away from me and pooped. It seemed directed at me, although that's probably giving myself too much credit. As if to compensate, another male Blackpoll Warbler foraged by. And then another. Seasons pass, birds come and go, may it play forward forever!

29. Hunters And Wildlife Lovers Alike Need To Take The Offensive Against Outside Cats

Cabin Talk.—One of the first things my wife asked me on our first date was "You don't have cats, do you?" In fact, I did, I adopted two stray cats, which I named Kate and Genevieve, and they'd been with me for 5 years. Her question was motivated because, you guessed it, she's allergic. Much to her dismay, Kate lived to be 18, and as most people know, cats quickly recognize who doesn't appreciate them, and they spend as much time as possible near these people. So, I don't have a "thing" about cats, as long as they're indoors and not outside (unless on a leash).

There are no scientific studies that conclude that it is a good thing to let house cats outdoors to kill native wildlife. The numbers to the contrary are staggering. Not wanting to quibble about exact numbers let's just say it's in the millions a day, and not less. This is simply unacceptable. In spite of the fact that these numbers are not debatable, the studies showing the detrimental effects of cats seem to have had little effect on the behavior of many cat owners.

My ire was raised by one of the more ignorant statements posted by an unrepentant cat owner on a site devoted to MN wildlife. Here are a few of the choicest.

"… it is not right to lecture anyone who lets a well fed, spayed/neutered cat out because "cats kill 3.5 million birds."" Fact one. Well-fed cats kill native wildlife, it is irrelevant that they are well fed, they kill irrespective of hunger level and do not belong in our ecosystem. Fact two. Spayed neutered cats suddenly do not stop eating or killing. Third fact. You can't have read anything serious on the topic if you say "3.5 million birds" unless you say "3.5 million birds a day."

"We can still love birds and the natural world at

Image Leidenschaft für Pferde und Liebe zur Natur, rihaij

the same time." I guess so, as long as you're ok with a lot less of them. And what about the "love." I guess cat lovers have a vicarious pleasure in watching their cat kill a naïve migrant warbler in its first fall migration. A bird that was hatched in the forests of northern Canada and making its way to Central or South America for the winter. The evolved genetic program would have gotten it there, but the same program did not evolve to teach them to avoid an introduced exotic predator that shouldn't be in the eco-system. These birds have enough problems: new preda-tors, food supplies, competitors, weather and other issues, and the last thing they need is a "well fed neutered cat" waiting to kill them.

"… there are feral cats colonies that live on res-taurant scraps [sic] in other places." There is nothing sen-sible here. Rats, roaches and flies live off the same scraps. And, while you might not see one of these French restau-rant cats kill a native bird, there's a reason – they've al-ready killed off the local resident birds.

"My cats like to go out, roll in the dirt, climb trees, and eat grass. I am not going to deprive them of be-ing cats because a bunch of non-cat owners said, You should...." Yes, cats behave like cats. Our native wildlife did not evolve with a feline predator of that size and abil-ity in the ecosystem. If you let your cats out, why not just shoot all the birds in your yard, at least they'll die quickly and not while a cat tears them apart.

We use public funds to provide parks and habitat for native wildlife. Then we look the other way when irre-sponsible cat owners let their cats out to kill the very wild-life for which we just paid to provide homes. There is no sense to this. We need to hold cat owners accountable for the wildlife their cats kill.

Good fences make good neighbors. This includes keeping your cat off my property or ANY public land that my tax dollars support. In most cases, you will know only a small percentage of the wildlife that your cat kills, even on your own property.

Here are my two suggestions:

1. All municipalities need to add "and cats" to the existing ordinances for dogs. If your cat is on the loose you'll be cited and fined.

2. Any cat on your property without your permission loses all legal rights.

It is currently a felony in Minnesota to kill a cat, even if it is feral – even if it is in the process of killing native wildlife on your property. Take about invasion of property! It is legal to shoot a pigeon, house sparrow or European starling on your property (or anywhere you have permission to discharge a firearm or air rifle or bow) because they are introduced exotic species. A house cat is the same as these species, it is an introduced, exotic predator that does not belong in the ecosystem. Why can someone else's cat violate my rights to a tranquil wildlife safe landscape? I demand the right to euthanize a cat on my property.

30. Bird Flu New And Old

Cabin Talk.—Most of what we see about nature on television and social media is warm and cuddly. Rarely do we see what nature really is, the daily life and death struggle between predator and prey, host and parasite, or diseases and populations. Life and death is truly part of nature, although some try to deny the death part. I typically have a relatively hardened perspective on death in nature. Not much in nature gets to me, with a couple old and one new exception.

I don't celebrate the deaths of deer, hogs, grouse, woodcock, ducks, geese, wild turkeys, among others, that I have legally harvested (hunted if you prefer) because they go on the dinner table (which I prefer to meat from commercial sources). However, there are exceptions to my hardened viewpoints. Images of loons dying from ingesting lead fishing tackle bring a tear to my eye. Hearing from vet techs about the mournful sounds made by lead poisoned eagles is gut wrenching. So much so that I have recycled all of my lead fishing tackle and replaced it with non-toxic alternatives; the cost was low compared to my boat, gas, yearly licenses, bait, tackle, etc. I fail to see why tackle companies have not seized on the sizeable fortune to be made in replacing lead with nontoxic alternatives. When you realize that lead fishing tackle is classified as hazardous waste, shouldn't that be a message?

Some discount the negative effects of lead on people. A scientific paper in the Proceedings of the National

Intense concentrations of snow geese provide fertile grounds for disease spread, such as bird flu. Image Bryan Hanson

Academy of Sciences found that over 170,000,000 Americans who were adults in 2015 were exposed to harmful levels of lead in their childhood, mostly from leaded gasoline. We have done much to correct this, but most reading this today were affected.

Another topic I find gut wrenching occurred in March 2022, and that is a die off of snow geese owing to bird flu, known as H5N1. I was sent a video of a snow goose that had lost motor control and was basically flopping around on the ground within a few feet of the camera (see image). I've never felt sorrier for a bird. This bird was not alone; it was one of many dying snow geese. Bird flu caused the deaths of thousands of snow geese in the mid-continent flock from Missouri to North Dakota.

H5N1 was big in the news about 15 years ago. We know what it is genetically, as the entire sequence of the influenza genome is as well-known as COVID-19. A common human form of influenza is H3N2. The H and N denote proteins (hemagglutinin, neuraminidase) on the surface of the virus that function in binding the virus to your cells; the numeric codes specify the type or strain of influenza.

The snow goose population burgeoned because large amounts of spilled grain on the wintering grounds resulted in high levels of overwinter survival. There is controversy, however, about how many geese there actually were. Some estimated that the midcontinent population numbered 3 to 8 million, a large range, indicating uncertainty. Some think that as many as 18 million were present in 2013. The consequence was that biologists believed that the geese were systematically destroying arctic coastal salt marsh habitat as well as adjacent freshwater habitats. How were they destroying it? They don't just graze leaves, they tend to pull entire plants out of the soil, which turns tundra into mud flats. Recent years have seen a reduction in reproductive success, with fewer young birds joining the population.

The USFWS instituted the Spring Light Goose Conservation Order (SCO) that allowed hunters liberal access to hunt snow geese migrating from the wintering to the breeding grounds. The hope was that hunters could reduce numbers by about 50% for the mid-continent population, thereby reducing ecological damage to the fragile arctic breeding grounds. That did not happen, and snow goose populations continued to rise despite spring hunting.

Snow geese are hunted for 9 months of the year, some birds are 15+ years old, and as a result in most flocks with adults there are birds that have seen every de-

Snow goose dying from H5N1 in South Dakota, March 2022. Photograph by Adam Toboyek.

coy, decoy spread and heard every caller. There are suggestions that migrating geese avoid areas with high hunting pressure. I've been in layout blinds and watched large flocks approaching our decoy spread only to have a few smart leaders lead the flock away at the last minute. I know decoys still work but the birds have become pretty savvy.

Despite controversy over whether the geese are actually destroying tundra, how many geese there are, the effects of the SCO, or reduction in reproductive success on the breeding grounds, population management might be taken from the hunters and in its place, H5N1 will reduce the population. H5N1 is carried by, but usually doesn't result in dire effects in, wild birds. It is, however, deadly to poultry.

What is news about the recent H5N1 outbreaks in the US is that wild snow geese and Canada geese are getting sick and dying. It appears that the virus has acquired mutations to make it also deadly to wild birds, not just captive poultry. That is, many wild birds have gone from carriers to being sickened by the virus. It is possible that it is spreading through wild migrating goose populations because at this time of the year they are especially concentrated, the condition that fosters spread in poultry. Because it is spread from goose to goose, the rate of transmission goes sky high when hundreds of thousands of geese are concentrated. The elephant in the room is whether people can contract bird flu from infected snow geese. Hopefully not, but I'm not positive. The spread to wild birds is somewhat concentrated in waterfowl, and birds that eat the carcasses.

Why should people be worried about "bird" flu? Simple. In areas of southeast Asia where it was common 15 years ago, the human death rate was about 50%, not less than 3% like COVID-19. It's deadly in humans because the fatal symptoms come on quickly, in a couple of days. Fortunately, despite its high death toll in humans, it appears, at least currently, very difficult to spread from birds to people, or person to person. The people that contracted it were basically living in and around infected poultry, with little access to medical care. The hope is that it will not acquire mutations that would make it contagious among people like COVID-19 or the annual flu.

It is common knowledge that wild populations can be controlled by disease. Thus, it is possible that influenza, specifically H5N1, might do what hunters have been unable to do, and that is to reduce greatly the numbers of light geese. It's nature's way after all.

31. Do Fish Eat Cormorants?

Cabin Talk.—Thinking we know something and actually knowing it often become mixed up. In this day and age, social media amplifies and give credence to ideas that are little more than rumors, and they explode and take

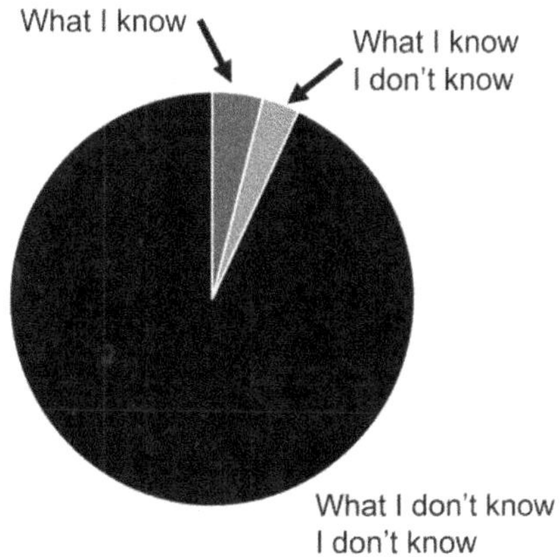

on a life of their own ("goes viral"). The diagram below serves as a reminder to all of us to take to heart what the Roman emperor Marcus Aurelius said, "The opinion of 10,000 men is of no value if none of them know anything about the subject."

Interests of people often conflict with wildlife. There is a large lake in northern Minnesota, Leech Lake, where a crash in the walleye population was thought to be due to the recolonization of the lake by double-crested cormorants. Walleyes are the most sought after fish in most of the Midwest, and I for one, count myself among the many who ply the waters for this amazing eating fish. An article stated that "Studies at Leech Lake have shown cormorants kill a high number of 1 and 2 year old walleyes" and that "Looking back, it seems clear double crested cormorants were the primary reason for the lake's fall from grace." Furthermore, DNR fisheries supervisor Doug Schultz, based in Walker, stated that "the 2001 year class was the one that disappeared on us from the main lake. I think that was direct predation." It seems clear, then, that cormorants are fish-depleting scoundrels and deserve relentless persecution. If you want to make a Minnesotan see red, take away his walleyes, or his deer (or more specifically, deer with antlers). But are cormorants really the culprit?

Double-crested cormorants declined in North America owing to harassment and shooting by people, and the agricultural use of DDT. Subsequent protection under the U.S. Migratory Bird Treaty Act of 1918, publicity by the Audubon Society, and availability of food at commercial aquaculture sites, led to its general recovery. Double-crested cormorants were documented at Leech Lake as early as 1821, although it is likely they were present at least on and off for thousands of years (history doesn't begin with our recording of it). However, cormorants apparently did not breed at Leech Lake again until 1992, a nesting absence of 160 years. By 1998, recolonization had taken hold and by the spring of 2004 there were over 5,000 birds nesting on Little Pelican Island. It was estimated that 10,000 cormorants used Leech Lake in the fall of 2004.

In 2004, the DNR began a program of culling cormorants using sharpshooters at the breeding colony, with the aim (pun intended) of keeping the nesting population at 2000. Fortunately, a large sample of dead birds was preserved, and their stomach contents catalogued. It is difficult to observe in the field what kinds of fish cormorants actually eat, and access to stomach contents is the best way to determine which species make up the bulk of their diet.

What's in the stomach of cormorants is only a start at understanding their effects on the fishery. The important question is whether cormorants eat fish species in relation to their abundance, or whether they target specific

Double-crested cormorant. Image: Samuel Stone

species no matter how abundant they might be. This is an important distinction. Also, it is important to know if adults eat different species than what they feed to the nestlings, or if prey preference varies throughout the year. For example, if cormorants don't eat, or feed their young, walleyes in the breeding season, but they do eat them in the fall, it would be undocumented by the breeding season diet data.

Birds were shot on their way back to the nesting colony to maximize the chance of identifying prey remains. The researchers kept track of whether the birds were subadult or adult, and whether they were males or females. In addition, regurgitated meals from young in the nest were put in ziplock bags (incidentally, that's some stinky business). Stomach contents were preserved and later identified. When possible, the lengths of the prey species were recorded. To ascertain if adults were feeding young a different subset of prey species, chick regurgitant was compared to the adult stomachs. That is, adults need to eat as well, and maybe some of the fish they catch they keep for themselves.

From 2004 to 2006, diet data were available for 716 adult cormorants, which included 34,900 prey items representing 29 fish species and two invertebrates. Both the percent of each species and their biomass was considered. This is important because a prey species might be represented by lots of small individuals or a few bigger ones. A big fish is still one fish.

The results of the analysis of cormorant stomachs were reported by Peter Hundt and colleagues from the U of Minnesota's Bell Museum and Conservation Biology Program, and the MN DNR. I was eager to see whether the cormorants were in fact walleye hogs. From the table, it is clear that walleyes are not a frequent prey item of breeding cormorants. Yellow perch were a major part of the diet in 2004 and 2005 but dropped off in 2006. It is apparent that there is a lot of variability from year to year. For example in 2006, whitefish, although not eaten in large numbers, constituted a large percentage of the biomass, meaning that fewer larger fish were taken (they have a high caloric value). Looking at the chart it is apparent that if diet data were available only for single year, it would be misleading.

A difficulty in interpreting these data is that we do not know the relative abundance of prey species that were available to cormorants. Hundt and colleagues note that cormorants are "opportunistic" feeders and suggest that they take the most abundant prey available. However, knowing this would require census data on the size distribution of prey that cormorants had to choose from (the menu) at the time the cormorant diet samples were collected. That is, you would rank the prey species based on their abundance in the lake and compare that ranking to what was recovered from the stomachs and young. If the rankings were the same, then yes cormorants are taking prey in proportion to their abundance.

Maybe the 2006 shift to larger whitefish resulted because there were too few perch around. Or maybe in 2004 there were too few large whitefish available. Without knowledge of prey species abundance, it is not possible to know how choosy the cormorants were being. It was interesting that Hundt and colleagues found that the young were fed a diet rich in yellow perch in both years, and more shiners in one year and more whitefish in 2006 (as in the adults).

Nature is a tangled web that is hard to decipher. Although there is no evidence that cormorants were taking a lot of walleyes between 2004 and 2006, either numerically or in terms of biomass, perhaps cormorant predation on yellow perch reduced the prey base that walleyes use and cormorants might have therefore indirectly reduced the walleye population. Possibly prior to 2004, cormorants were taking larger number of small walleyes when they were more abundant and this might explain the "loss" of the 2001 year class. However, during that period there were relatively few (around 1700) cormorants nesting and none were recorded in the fall. And today, there are about 2000 nesting cormorants and walleyes are increasing. Also Albert noted that from 2005 to 2013 that the DNR stocked over 140,000,000 walleye fry in Leech Lake, and this could be a factor in the upturn in walleye numbers as well. I think that we have yet to understand fully the role that cormorants played in the decline in perch and walleyes in Leech Lake. The data from 2004 to 2006 suggest it's a lot more than double-crested cormorants.

Main prey species in diets of adult double crested cormorants at Leech Lake from 2004 to 2006 (reported by Hundt and colleagues).

Species	2004		2005		2006	
	% prey	%biomass	% prey	%biomass	% prey	%biomass
Shiners	18%	12%	19%	6%	52%	8%
Yellow perch	79%	71%	75%	78%	34%	37%
Whitefish	<0.01%	0.1%	0.4%	3.7%	2.5%	43.4%
Walleye	<0.01%	8.1%	0.2%	2.2%	0.7%	3.0%

Graphs can be dangerous, as they show correlations, not necessarily cause and effect relationships. Comparisons of population trends of two species can be interpreted in several ways. For example, the chart that Albert showed reveals that cormorants decreased in recent years, and the DNR has found that fish populations increased. Now tongue in cheek, one might infer from this relationship that the now abundant fish are eating cormorants resulting in their decline. Then it would be appropriate to suggest culling fish!

32. Never be a baby bird

Cabin Talk.—Few things in nature jerk the heartstrings like baby animals. The raging mother bear protecting her cubs, a doe fending off a coyote attempting to catch her fawn, a bird attacking a snake about to devour her nestlings, even male bass guarding their nests against marauding panfish. All these evoke deep emotions in people, me included. Who doesn't cheer for the protection of parents provide to their young, even when it's life threatening to them?

I learned ornithology from Dr. Dwain Warner, a long time curator of birds at the University of Minnesota's Bell Museum. Dwain, or DW as we called him, had a number of sayings, but one of his favorites was "Never be a baby bird." He was referring to the fact that the fate of most baby birds is to perish before leaving the nest, shortly after fledging, or even hatching from the egg. For example, in species like the black-capped chickadee, something like 90% never make it to their first breeding season. Hence, DW's admonition to avoid being a baby bird, and I keep up the tradition.

Baby birds are extremely vulnerable, especially those that we label as altricial. That is, the young are hatched essentially featherless, eyes closed, and basically 100% dependent on the adults. Young altricial nestlings are essentially living guts with a head and legs. When a predator attacks, typically the entire nest is wiped out. My field ornithology class at Lake Itasca and I once watched a red squirrel go into a nest hole in a dead tree and bring out all of the baby black-capped chickadees one by one, eating them like they were ears of corn, partially grown feathers raining down. All the while the adults were scolding the squirrel, who wasn't bothered in the least. The squirrel ate them all.

The other extreme way to be a baby bird is to be precocial. Species like ducks or turkeys have babies that are basically ready to go right out of the box (egg). They can see, hear, follow mom (imprint), run and hide, and mostly feed themselves after a short period. If you've followed a group of Canada Goose goslings in the spring on a local pond, you notice that, unlike the altricial birds, the little flotilla seems to decrease slowly over time – sometimes one less gosling a day. By being able to scatter when the parent sounds the alarm, predators typically don't get the whole bunch at once but pick them off one by one.

But in the end, DW was still right. We are not neck deep in birds and the reason is that it's really tough to make it to adulthood, whether altricial or precocial. Predation prevents populations from growing out of control. Now, I can think of a bunch of predators that would take baby goslings or ducklings, but Lisa Dessborn and colleagues asked a slightly different question. They wondered not which predators would take baby mallards, but how do mallard ducklings react when confronted with potential predators? Do they have some built in ideas of what to avoid because it might want to eat them? Part of the motivation of their study was that you'd have to spend a ton of time to see many predatory attacks on ducklings, so why not ask the mallard ducklings who they're afraid of? The answers were to be found in Dessborn and colleague's paper in the journal Behaviour.

Mallard hen and chicks. Nennieinszweidrei

The authors bought mallard ducklings and introduced them to a series of fenced enclosures surrounding small ponds. They created broods by keeping birds together (a bit of paint on their backs helped the researchers know who belonged with whom). They devised ways of scoring the duckling's responses to various threats.

They exposed the ducklings to various threats. First, there were calls from crows and gulls, known duckling predators, and for comparison, they used calls of native species, like finches, which are not threats. Two dead pike were purchased from a local fisherman (says something about the researcher's fishing skills?). The broods were removed, and the (frozen) pike was placed in the center of the pond. They also simulated a pike attack by attaching the dead pike to a fishing line and rapidly moving it towards the brood. To simulate aerial attacks, they used a stuffed goshawk that 'flew' over the ponds along a wire. Unfortunately, there were no "controls" in either of the last experiments. The researchers ought to have flown something like an oriole or a milk bottle over the pond, and used a non-threatening aquatic creature, or a bottle of water that had sunk to the bottom.

Lastly, because the introduced American mink are huge duckling predators, a captured mink (that had killed a brood and parts of two others) was put in a cage next to the pond, in view of the ducklings.

I was really intrigued by this. What would the ducklings have hard wired in terms of predator avoidance? When they heard the recorded bird calls, they swam to the middle of the pond and craned their necks, with "apparent vigilance." However, the response was much stronger to the predatory birds, showing an innate response to a potential aerial attack. When the birds saw the goshawk coming towards them, most but not all of the broods dove and scattered and surfaced some distance away. I was surprised the response was not stronger. They also noticed that some ducklings seemed to be good followers, reacting to other's behaviors.

Oddly, the birds were not terribly concerned about the mink in the cage, although they did tend to stay on the other side of the pond.

The simulated pike attack, however, was a different story: "As the pike dummy broke the water surface, ducklings in all broods scattered about by running on the water surface and onto shallow waters, and two ducklings in one brood even ran up on land." Now that meshes with a similar attack I saw on a brood in northern Minnesota — chaos in the water. The broods were apparently nonplussed by the suspended pike. This was true even after the broods had been "attacked" by the pike. Maybe they "sensed" that it was frozen, and I don't take much from this observation. Of course, the alternative, a live pike, might have ended their experiment.

The results were not as straight forward as I thought they'd be. I was most "bothered" by the lack of response to a motionless pike, apparently sitting in ambush. So were the authors, as they pointed out that pike are widespread, they eat ducks and ducklings, and ducks do not avoid lakes that contain pike. The authors suggested that ducklings are unable to identify pike and other large fish or even similar shaped objects apart when under the surface. They likely rely on mother mallard for this. I personally would have seen if they found a snapping turtle decoy similarly unthreatening.

There is an obvious tradeoff here, in that ducklings cannot avoid water; in fact, we think it helps them avoid land based predators. Furthermore, they feed on the water's surface, and they can't just not eat to avoid being eaten themselves. So maybe part of the strategy is waiting until a clear threat emerges, like a large fish heading towards you before scattering. And we should not overlook the fact that "scattering" is likely a strategy to confuse a predator, potentially causing a moment's hesitation on the predator's part, resulting in everyone's escape. I subsequently saw a home video on YouTube that showed this exact result – a northern pike in Alaska attacking a mallard brood from underwater.

The authors also commented on the unexpected result of little response to the mink in the cage. They pointed out that ducks in Europe have not evolved with a predator of this kind, and perhaps local ducklings lack the appropriate response. That I doubt, but I don't have a good explanation either. Someone ought to import some eggs from North America and do the same experiment.

In the end, DW's pronouncement still rings true. Even in captivity, a mink broke in and ate a bunch of baby birds. So, altricial, precocial, or in between, never be a baby bird.

33. Loon Hunting: thankfully a bygone tradition

Cabin Talk.—"On 6 May 1950, eleven federal and state wildlife enforcement officers staged a long planned, coordinated raid on Shackleford Banks, Carteret County, North Carolina, apprehending nearly 100 hunters, of whom 78 were formally charged with illegally shooting loons (Anonymous 1950) in violation of the Migratory Bird Treaty Act of 1918." So begins a fascinating article on the hunting of common loons written by Storrs Olson, Horace Loftin, and Steve Goodwin in the Wilson Journal of Ornithology. I had never heard of loon hunting, but what I learned was an astounding look at the early history of our state bird and its past perils on its east coast wintering grounds.

The Shackleford Banks (see map) is a narrow east west barrier island that separates the open Atlantic on its south shore from Harkers Island to the north, with Back

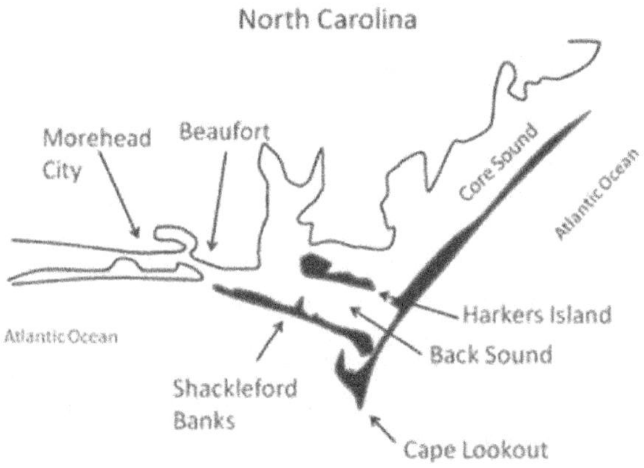

Map showing where loon hunting occurred.

Sound in between. Shackleford Banks was settled by the early 1700s by people who made their living catching whales, mullet and dolphins. The area was devastated by the San Ciriaco hurricane of 1899, and resulted in people moving inland, many to Harkers Island. The banks were mostly uninhabited and used for grazing cattle; they became part of the Cape Lookout National Seashore in 1966.

Loons are fairly large birds, with males averaging 16 lbs and females 10 lbs. They were never a common species in the bags of waterfowlers or market hunters, because they are hard to hunt and have a reputation as being too "fishy" to eat. They do not respond well to decoys. Although in winter they molt their flight feathers offshore and are flightless, it was too hard to pursue them with boats because, as most of us know, when they dive it is hard to predict where they'll come up, and they often ride low in the water, making them a difficult target. As Minnesotans know, they require a running start across the water's surface to become airborne. When they are migrating, they often fly very high and well off shore, out of shotgun range. And, although it has nothing to do with this article, close your eyes, and recall their eerie vocalizations that

Harkers Island resident Otis C. Willis (b. November 1918, d. June 1996) with the results of a loon hunt in 1943.

fascinate us during their summer residency on northern lakes.

If loons are hard to hunt and taste fishy, how could there be a tradition of loon hunting on Shackleford Banks? There was a flaw in the loon's migration strategy. Along Shackleford Banks loons become common in spring and take off to the north from just offshore to begin their northward migration, bringing them low over the banks and within shotgun range. The original Shacklefordians took advantage of this to put meat, albeit loon, on their tables. Harkers Island became the focal point for this activity, although there were other areas where loons were killed in large numbers. By the mid-1800s the hunt was well established.

Generally in April and May, hunters would depart Harkers Island before dawn, especially with a northeast wind, and beach their boats on the north (protected) shore of Shackleford Banks. They walked southwards across the dunes to await the first flights of loons that occurred at first light. Hunters didn't require much concealment although during WW II they sometimes hid behind anti-submarine buoys that had broken loose and beached. At times there were so many hunters that they were either stacked up behind one another, or they were at least spread out just out of shotgun range of each other; there was often no place a loon could fly over without being in range of a gun. One resident of Harkers Island recalled that "shooting began so regularly at daybreak that it was as good an alarm clock as you could ask for" and that at one point during World War II "the shooting sounded like an invasion."

The loons were taken back without being field dressed (see photo). This doesn't seem particularly odd, but one hunter might shoot over 15 loons, which at 10+ lbs each, amounted to quite a load to drag back to the boats.

I kind of figured that this was more about shooting sport than food. Instead of sporting clays like we shoot today, "sporting loons." However, the residents vigorously defend loons as good table fare. They even claimed that they preferred them to ducks and geese, and that to residents of Harkers Island the fishy taste was "barely discernible." One cookbook refers to loons as "Harkers Island turkey." Olson and colleagues noted that "For 19th century residents of Shackleford Banks who had survived a winter on salt fish, root crops, and such grits and flour as they may have received in trade for mullet and whale oil in the previous months, a loon represented a sizeable chunk of fresh meat as well as a welcome change in diet." A bird not consumed by the hunter might bring 50 cents on Harkers Island.

A second use of loons was to make the bones into fishing lures, which were used to catch bluefish and Spanish mackerel. Bones were sold to local hardware stores for 10 cents. The bones were bleached and slid over a hook, and the white coloration was apparently attractive to fish.

It became illegal to kills loons in 1918 with the passing of the Migratory Bird Treaty Act. The hunting continued in that local area for quite a while, and during the second war, there were other things to worry about. But the hunt became too popular, and the locals were joined by others from some distances away. This was too much for law enforcement to ignore, and in the 1940s an undercover agent settled locally and began documenting the loon hunts and planning the raid mentioned above.

It was estimated that the hunters had killed 200 loons on the morning of the big raid. Hunters who were apprehended were fined $25 each, which is about $225 in today's dollars. The raid was deemed a big success for law enforcement as it basically put an end to the illegal loon hunts. Shortly after the big raid of May 1950, residents reacted negatively, suggesting that "laws that overregulated hunting had come into effect because of "fancy" "upstate" sportsmen, whereas the Carteret County loon shooters participated in the "only exciting sport left untouched by the law. Their father had done it, their grandfather before, even before their forefather had moved from the outer banks. You weren't a man until you had shot a loon."

I don't bear a grudge against the loon hunters, as I think you have to understand the context in which they participated. It is thought that there are still a few loons killed today that secretly end up in local stews. It's hard for me, not having been part of the local culture at that time and place, to pass judgment from afar. I'm glad to have read about the tradition, and glad it's ended. Nothing adds more to my northern Minnesota experience than the sounds of loons. And although I'll eat just about anything, I'll pass on loon.

34. Long term sexual tensions between male and female ducks

Cabin Talk.—Once people learn I'm an ornithologist, I sometimes get motioned into a corner where I'm asked in a hushed tone, "say, how do birds 'do it'?" Birds are at the opposite end of things like some worms, where "the act" can take seven hours. In birds, the act is not particularly dramatic, at least by Hollywood standards. But the actual act itself is not well known among many people. Even I, a professional ornithologist, was almost shocked by the discoveries I'm presenting here. And fyi, this article was rejected by a newspaper because the word "penis" appeared too many times.

Most birds do not have a penis. Copulation is usually little more than a brief meeting of genital openings, which are termed the cloaca. Now, granted, some birds like Lapland Longspurs copulate 350 times per clutch (4-5 eggs), so there can be a lot of action.

And courtship in birds can be a prolonged, expensive affair. Some males bring the female gifts of food, our version of taking her out to dinner, but actually, more likely illustrating the male's ability at finding quality food that will be crucial when bringing up baby. And some pairs go dancing, like the Western Grebe, where the pair race across the water's surface in perfect choreographed unison, what David Attenborough called a pas de deux (a dance duet). Males of many species show off their value by growing showy plumages, often to the extreme.

Take the peacock. Not even a staunch adaptationist would suggest that the train of the male is for camouflage, tricking potential prey items, or in aiding in aerodynamics. Most think it exists to impress the female, pure and simple. But how? There is evidence that the "eyespots" are an indicator to the female that the male is of high genetic quality, and in particular that he has resistance to disease or parasites. So to impress females more so than rivals, males evolve ever greater numbers of brighter eyespots, females choose them preferentially, and the process escalates in what evolutionary biologists call an "arms race." This process was termed sexual selection by Darwin (he got an impressive array of things right), and it occurs via the process now known as "female choice."

But let's return to the basics of bird reproduction. I mentioned that few birds have a penis. In the past few years, our understanding of this part of the bird's anatomy has exploded. It turns out that many ducks have a penis that is very long (up to the length of the male's body), twisted into a spiral, grown in the breeding season and lost shortly thereafter (explaining why hunters don't notice them in the fall). The Peking Duck is a particularly good example (see photo).

These spiral shaped penises function differently than most vertebrates. At rest, the penis is kept inverted (i.e., outside in). The penis literally explodes from the base under pressure from the lymph system (not blood supply like in mammals). This explosion, taking 1/3 of a second, only occurs when the male and female genital openings are pressed together. That is, the erection (called an eversion by duck researchers) does not occur before mating. Sperm does not travel inside the penis, but in a groove along the external surface.

Now, we can ask what the penis of ducks has to do with eyespots on the tail of a peacock. To answer this question, we first have to consider why some ducks would have a penis that is spiral shaped and dramatically longer than "necessary," when it would seem a simpler design would do? That is, like eyespots on the peacocks train, what's up with the penis of ducks (pardon the pun)? Could there be a role for female choice?

Many have seen the aggressive behavior of mallards in the spring. Two or more males are often seen in high speed pursuits of a hen. One of the males is likely her mate, but the objective of the others is well understood, "forced copulation" (FC). If you see such a chase, it is obvious that the hen is trying to escape, as she actively tries to ditch her pursuers. Often, however, the males force the female to the ground or water and forcibly attempt to mate with her. On some occasions, the frenzy has resulted in the drowning of the hen mallard, and the same has happened on land.

Observations suggest that the female is not actively interested in mating with these other males. However, paternity studies have shown that it is an effective strategy on the part of males, as a proportion of mallard hatchlings are fathered as a result of these FCs. Still, our theory is clear that females ought to be choosy about whom they mate with, and these flights are an obvious way to attempt to avoid unwanted pairings. But hens sometimes do not escape. Is there some other way to enforce "birth control'?

Back, again, to duck reproductive anatomy. Current thinking is that it's not just the male anatomy that is odd, but the females as well - the vaginal pathway is convoluted and sometimes has dead ends. Some researchers

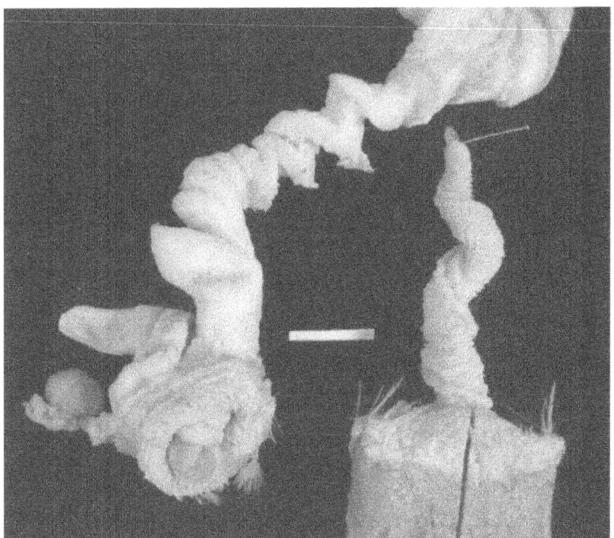

Male (left) and female reproductive tracts from the Peking duck, showing the co-evolved shapes of the penis and vaginal tract. Photo courtesy of Dr. Patricia Brennan.

speculated that this could be a way that females prevent unwelcome inseminations. That is, instead of a lengthy and potentially dangerous flight to avoid an amorous male, perhaps her anatomy can help thwart unwanted matings. To know this, we have to know the intimate details.

What needed to be discovered was not going to come from observation, because bird sex, the anatomical part, is pretty discrete. Some researchers decided to see if there were any consequences of the male's odd-shaped penis, by experimenting! Dr. Patricia Brennen from the University of Massachusetts and her colleagues set up several devices designed to mimic a female duck's reproductive tract. There were straight tubes, spirals with a clockwise and counterclockwise turns, and one with a 130° turn, which mimics the female anatomy. Although they might seem like weird attempts, these were brilliant experiments!

They found that males were successful in everting their penises in the straight or clockwise spirals, but not with the others. That is, the replica of the female's reproductive anatomy could prevent the explosive extrusion of the penis. But, you say, if they weren't successful with the device mimicking the natural female, why are there any ducks left? Here's where the analogy between the escape flights of hen mallards and the reproductive tracts of ducks come into play. When mating, females of species in which forced copulations occur noticeably change the position of their body during mating, depending on whether it is a wanted mating (i.e., with her chosen mate) or a forced attempt. If it is a forced attempt, they can prevent full eversion by assuming a particular position, and the sperm is deposited too low in the reproductive tract for fertilization. But when it's the correct partner, the hen assumes a position that allows full eversion and successful insemination. So, the researchers perhaps failed in that their apparatus didn't mimic a receptive hen, but instead an unwilling participant.

The researchers speculated that the reproductive anatomies of males and females have undergone an "arms race" and co-evolved. To prevent unwanted matings, female anatomy evolves a new twist, so to speak. To counter, male penis shape co-evolves to be more successful, and so forth. What is left is a reproductive system that is chock full of moves and counter-moves, and is different from what one would speculate is "necessary." Very much analogous to the eyespots on the peacocks train. Females are in essence choosing particular aspects of male anatomy, and they evolve in response, just like Darwin predicted.

Not all ducks have this elaborate co-evolved reproductive anatomy. In species that do not have forced copulations, there is no elaboration. But there are some new "twists" on the story. Evidence suggests that some male ducks actually grow longer penises when there are more males around, making them the most likely one to father the ducklings. Once again, nature has been hiding some spectacular details from us. One of my ornithology students even suggested that my lecture on this topic should be entitled "Pornithology."

As a postscript, in spring 2013 Brennan's research was attacked in the media by politicians that did not see the value in her research. Obviously, it's easy to make fun and challenge the need for federally funded studies of duck penises. However, the criticisms amounted to an ignorant mockery of science, often seen in such sources as Fox News. Her work, funded by the National Science Foundation, passed a panel of experts who evaluated the scientific potential of her work, and she was awarded a grant. Given that funding levels are less than 10%, this means the experts found a great deal of importance to the research.

Not all research has to directly benefit humans. If you stifle the scientific process because you do not understand the research, we all suffer. There are hundreds of instances in which a discovery in one field helped to remove a roadblock to progress in another. Granted, not all research is deemed important by other experts, and one way this is accomplished is via the funding process. To say that a piece of research has little value ought to be done by persons qualified to judge. In Brennan's case, her research was extremely insightful and valuable to the field of animal behavior, and her critics badly misinformed.

35. How much wood could a woodpecker peck, before getting a headache?

Cabin Talk.—Woodpeckers are great, except for those that take a liking to the wood siding on your cabin. I love hearing the sound of a yellow-bellied sapsucker drumming on the metal lid of my neighbor's garbage can just after dawn. The drumming starts out fast and then tapers off to a slow beat then ends. It's the male sapsucker's way of telling other males to buzz off, and any females within earshot that he's available. Once you've heard this drumming coming from a metal garbage can lid, you might ponder how this could possibly be adaptive, it's got to be a harder surface than a tree? It might be like us chewing cement.

A common reaction we all have when we observe animal behavior is to put it in the context of us doing the same thing. If you stick a fish hook in your lip, you feel pain (and frantically hope it was the barbless one). Does a perch with a hook in its mouth feel the same kind of pain you do? Scientists are very divided on this question, with perhaps a majority saying that fish lack the pain receptors (nociceptors) that would result in their feeling pain the way you and I do. Instead fish feel some sort of "discomfort" but not the way you and I feel pain.

Speaking of fish hooks reminds me of a time I took an older friend fishing, and we were casting for northern pike on a lake in Northern MN. She noticed that I winced on her backhand, but she scolded me and assured me she was a pro at casting. Not two casts later I feel a clunk on my backside. Yup, she had gone too far in her windup and impaled the hooks from a large spoon in my back pants pocket. Fortunately, it was the pocket I kept my wallet in, so I didn't get to discover what being hooked feels like. In a local resort, the ceiling is covered with dollar bills, which after inquiring, I learned that each was from an angler from whom resort personnel removed an embedded hook.

I'm sure just about everyone has observed a woodpecker going about its business of drilling into trees in search of insect larvae or making a nest or roost hole. The "drumming" is also an auditory signal that serves to advertise for a mate, or to warn other males that this spot is taken. Woodpeckers don't just give a halfhearted peck or two. A bird can peck at a speed of up to 23 feet per second, 20 times a second and up to 12,000 times a day. That's the stuff of which headaches and concussions are made.

Woodpeckers have evolved some cool adaptations. First, the tail feathers are very stiff and strong, allowing the bird to prop itself upright on a tree trunk. It has strong feet, with two toes facing forward, two back ("zygodactyl" in case you wanted to know). Perhaps most interesting, they have a long tongue that curls up under the skull, around the back, and fits in a groove on the top of the head! They use the long tongue to extract insects from the holes they drill, and they had to figure out some place to put it when it wasn't in use. Anteaters have nothing on woodpeckers.

Helping the woodpecker are a couple of other adaptations that are not woodpecker specific. First, birds have a bone in the tongue called the hyoid apparatus, which provides support. A long skinny tongue needs a little help when probing in holes. Second, birds have a bone in the eye called the sclerotic ring. Because birds have relatively large eyes the bone provides support, one would imagine especially for woodpeckers when they're banging away on a tree, or your siding.

Back to pecking. Although we have a saying about banging one's head against the wall, watching a woodpecker drilling into a tree leads to the inescapable conclusion that you wouldn't want to mimic a woodpecker. But it seems straightforward to assume that woodpecker skulls must be equipped with shock absorbers. In fact, we've taught ornithology students for decades that a foamy layer between the birds' bill and skull absorb the impact, cushioning their brains. Some think the bones of the skull are specialized and that the tongue acts as a spring like shock absorber. How else could it be that you don't see a woodpecker in the concussion protocol?

Fortunately, someone finally set out to see if traditional wisdom was right. A scientific paper in Current Biology entitled "Woodpeckers minimize cranial absorption of shocks" by Sam van Wassenbergh and colleagues explored this question and its implications for people.

The National Football League has become serious about the long term effects of concussions on its players. A human is concussed when the force of the blow to the head is about 80g of force. Calculations of the force on a woodpecker brain is estimated at 1,000 g of impact. It would seem like woodpeckers would be a worthy model for the NFL.

The new paper makes a somewhat astonishing claim. They say that the woodpecker skull doesn't absorb shock, and that if it did it would be maladaptive - the woodpecker would have to peck harder to accomplish the task. How did they figure this out? They took some exceptional slow motion video of woodpeckers pecking (just search online) and tracked the movement of three spots they drew on the images, one near the bill tip, one in the middle, and the eye. They reasoned that when they measured the movement of the bill and eyes during the shock from a peck, if there was cushioning, the eye would lag behind the tip of the bill because of a built in shock absorbing device.

They found nothing of the sort. The heads are very stiff during pecking, and there was no shock deceleration of the braincase, the eye and other two points moved at the same rate. The conclusion is that a stiff hammer makes a better tool. To drive a nail into a board as efficiently as possible, hit it as hard as you can as few times as needed, while avoiding your thumb (which could be an unfortunate shock absorber). In woodpecker terms, the paper stated that the bird "would need to hit their selected spots twice as fast as observed or strike at its top speed on wood that is four times as stiff to suffer a concussion." Well, at least it's possible. It still seems remarkable that woodpeckers are not concussed with all that force, force greater than we anticipated because of the lack of shock absorbers.

There's another piece to the puzzle. Turns out woodpecker brain size is small relative to that needed to sustain concussions. In other words, the force imparted to the relatively small woodpecker brain is below the threshold that would cause a concussion in a primate. This might explain why no woodpeckers have won a Nobel Prize. The moral, then, is don't bang your head against a tree unless you're a woodpecker.

36. Wood ducks: not always nice, sometimes naughty

Cabin Talk.—Typically, papers appearing in mainstream scientific journals are laden with large sample sizes, lots of data and analyses, and sometimes long winded interpretations. I've authored a few myself. A paper appeared in a mainstream journal about wood ducks. Two wood ducks, that is. And some eggs from a single nest box. This sounded like blog material, not a scientific journal article. But there's more to the story. In spite of the actual small number of ducks and eggs in the study, what was learned was quite impressive.

Everyone knows that wood ducks natively nested in holes in trees in the woods (hence, the name) and will now also readily use nest boxes. Nest boxes are often parasitized by female wood ducks, as some hens are ready to lay an egg but lack a basket in which to put them. In fact, the small-bodied brood parasitic brown-headed cowbird has even dropped an egg in a wood duck nest (to the cowbird's detriment). On the whole, there are not a lot of birds that have adopted the brood parasitic lifestyle completely, but quite a few waterfowl like the wood duck do it opportunistically. In addition, if a hen already has enough eggs in her nest, why not drop one in someone else's nest? If you can get another female to incubate and care for one extra of your offspring, it's a win-win for you.

One might think it is pretty shady to put your eggs in someone else's nest. But the strategy can be very profitable. Female brown-headed cowbirds have dumped an egg in the nests of over 200 other species, some of which are just inappropriate (like the wood duck). Many species reject the eggs, but others raise the baby cowbirds, which usually results in the death of the host offspring because cowbirds hatch sooner and grow faster than the host young. If parents of acceptor species find a brown-headed cowbird, they can't seem to help feeding it constantly and often neglecting their own offspring.

Speaking of wood ducks. It is pretty difficult to monitor a wood duck house for an entire season and see all comings and goings, but there are other ways of figuring out what's been happening. Kayla Harvey from the State University of New York and the Delaware DNR and her colleagues published a paper in the scientific journal PlosOne on two hen wood ducks and their eggs. It turned out that the study was in fact pretty data rich!

As part of a study of 239 artificial wood duck nest boxes, it was observed that on 1 May 2020, two different females were incubating eggs in the same nest box. One of the females flew away when the technicians approached, the other sat tight. The big surprise came a few days later when on 4 May, one of the hens was incubating the eggs next to the other female, who was now deceased! Interestingly they both tried to use the same nest box the year before.

The living hen continued incubating until 6 May, when all but one egg hatched and the little fuzzballs did that baby wood duck death defying leap to the ground. The unhatched egg and all of the eggshells were collected, and the researchers put on their molecular detective hats.

You can extract DNA from the adults (via blood or tissue samples) and membranes from inside the eggshells. One of the molecular tools they used was mitochondrial (mtDNA). As in humans, you get all of your mtDNA from your mother (that is, inheritance of mtDNA

Ducks Unlimited Canada

is the opposite of male surnames). What the study reported was that at least four unrelated females had laid eggs in that nest box. The eggs in the box were not from a group of related females — that is, they were not sisters or all that closely related. This is news because they verified the number of egg-dumping hens (the nest box was probably too exposed but that's another story — just remember, they're called "wood" ducks for a reason and not open-country ducks).

What about the two hens in the nest box? The researchers wondered if they were close relatives or even sisters, which might explain the attempted joint occupancy. As you would guess from the observations on the parentage of the eggs, the two hens were also unrelated. There is no indication as to what caused the death of the incubating hen because the authors didn't do a thorough necropsy.

There was one other surprise from their molecular snooping. They found that a single female can lay eggs that were fathered by more than one male. The way they discovered this is pretty standard molecular forensics (unless you're from the LAPD). From the mtDNA profile and multiple profiles at the other genes surveyed (from the nuclear genome), it could be determined that eggs in the nest were from at least one female who had multiple male partners. In particular, it was found that the deceased female laid an egg before expiring that was a half sibling to the rest of her eggs. My oh my, what a tangled web.

Not to stray into duck morals, but one might ask why the female had multiple male partners? This phenomenon, termed extra-pair copulation, is a way females can increase the genetic diversity of their offspring, which presumably increases the number of the hen's young making it to adulthood. As discussed elsewhere we know that male and female ducks have complex, co-evolved genital systems, where the penis is basically a corkscrew that will only fit in the female should she be so inclined! Hence, it was probably a deliberate act on behalf of the deceased hen, although we're not entirely sure if females can always prevent these extra pair copulations.

The combination of a fortuitous find and some clever molecular forensics allowed Harvey and colleagues to glean a great amount of information from just two hen wood ducks and their eggs. We now understand even more about the love life and times of wood ducks.

37. The Cost of Raising Baby Birds: a Cowbird Perspective

Cabin Talk.—Most folks are aware of at least the common birds around their homes. One of the birds that you'll see in most places is the brown-headed cowbird. This bird defines an evolutionary cheater. After reading below, you might hate them or be impressed; in any event, they are federally protected! To nest or not to nest, that is the question birds face most every year. Obviously, most species go through the involved nesting process so that they can raise a brood of young.

There are significant costs typically involved in raising a baby bird: finding a mate, defending a territory (often only the male), having energy to make eggs, mating, building a nest, laying eggs, defending the eggs, incubating eggs, feeding nestlings, and caring for young after they fledge (at least they don't move home). If a bird could avoid paying some of these costs, it might be at an evolutionary advantage. There is such a strategy and it is called brood parasitism.

There are two basic types of brood parasitism, obligate and facultative. Obligate brood parasitism has arisen independently in 5 families of birds. In North America, only the brown-headed cowbird, named for the brown head (and black body) of the male (the female is basically drab brown all over) is an obligate brood parasite. Elsewhere it has evolved in indigo bird, honeyguides, black-headed duck, and Old World cuckoos. Several species of African honeyguides are especially bad if you're a host young, because the newly hatched, naked and blind honeyguides have "fangs" on the tips of their bills, which they use to kill host nestlings, thereby getting the host parents full attention. If you're interested in the gruesome

details,   check   out
https://www.youtube.com/watch?v=8j6Qu3H9oTY.

Some species are facultative brood parasites, birds that do their own nesting and dump a few eggs elsewhere for (their own) good measure. Species such as our black-

Chipping Sparrow eggs (blue) and brown-headed cowbird (larger egg on right). Image: The woodlot.

billed cuckoo and yellow-billed cuckoo come to mind, as well as several ducks. Making the jump to obligate parasites means finding and exploiting at least one host species where the parasite can be successful for the long haul. Of course, in either strategy, a goal is to be prudent and not run out of hosts by parasitizing them out of existence.

Female brown-headed cowbirds are sneaky and carefully watch the nesting efforts in their local areas. When they find an active nest of a potential host species at the right stage, just after the host has laid an egg or two, they come in when the host female is away and lay their own egg, with varying fates. Some birds are called acceptors, others rejectors. One rejector is the yellow warbler, which when confronted by a cowbird egg in its nest builds another nest on top of its previous nest, dooming the cowbird egg(s) as well as its own. The reason is clear, there's strong evolutionary pressure to avoid raising cowbird eggs at the expense of your own.

Other species accept cowbird eggs, which in many cases leads to the loss of their own young, because of adaptations of cowbirds. Some adaptations include a relatively rapid incubation period, meaning the cowbird young are already hatched and being fed by the host adults (the urge to feed a baby bird is strong!) before host eggs hatch, putting the smaller host nestlings at a disadvantage. Sometimes the cowbird baby is just bigger, also an advantage. Seeing a chipping sparrow feeding a fledged cowbird makes this clear. The acceptor species don't seem to notice that these baby cowbirds are not theirs, only that they are really big and hungry and they must be fed. You'd think that a chipping sparrow would wonder why it's baby looks and sounds so different than its own nestlings and is twice adult size (three times when they're done growing).

Brown-headed cowbirds eggs have been found in over 200 other species, some of which are just a matter of unsuccessful necessity, such as the egg found in a wood duck nest! Given how evolutionarily clever female cowbirds are this seems pretty odd. It is possible that the large number of attempted hosts is a function of the easterly expansion of cowbirds following the clearing of forests. This relatively new part of the range brought cowbirds into contact with species that had not (yet) been exposed to brood parasitism and the have not yet had time to evolve brood parasite defenses. It might take thousands of years or more to evolve an effective parasite defense.

Returning to costs of nesting, consider that a female bird the size of a brown-headed cowbird might lay a clutch of four eggs, and if they are lost to predators or weather (very common occurrences), can probably renest and lay another 4 eggs. A brown-headed cowbird female does not pay the "costs" of nesting, and this energy savings allows a female cowbird to lay up to 40 eggs in a season. Of course not all will survive, or we'd be neck deep in cowbirds. But it's a huge reproductive advantage (40 to 8) to the cowbird.

Here's a common question (for me too). How does a baby cowbird that was raised by host parents of a different species know it's a cowbird and not grow up and fall in love with, say, a sparrow like its parents? Adult cowbirds hang around nests that contain baby cowbirds, vocalizing, and when the cowbirds are independent, they join the flock that speaks their language, so to speak.

The evolutionary transition to brood parasitism is difficult, given its rarity. But if it can be accomplished the benefits to the parasites can be large.

38. Do Geese Get Scared?

Cabin Talk.—Throughout much of their ranges, different species of geese cause agricultural damage. For example, an older figure for western US noted that crop farms lose as much as $30,000 annually to geese. A Canadian researcher concluded that prairie wide losses due to waterfowl ranged up to $10,000,000 annually. As a consequence, wildlife management agencies open early goose seasons, and a spring hunt for white geese (Snows, Ross'). But, as far as the white geese are concerned, there is no clear effect, at least as yet.

In Europe, there are some clear figures for damage. For example, Swedish farmers have been paid up to 750,000 euros (about 815,600 US dollars) to compensate for crop damage caused by geese. In the Netherlands, up to 5 million euros have been paid, and in Germany, losses in hay harvest increased from 15% to 50% from 1996 to 2010. Thus, crop damage is not just a US problem.

One way to control goose reproduction is to reduce breeding, such as in Minnesota where the DNR puts oil on eggs to prevent hatching. The city of Duluth, Minnesota has a goose management program in which eggs are covered with corn oil and left in the nest, and the geese continue incubating but the eggs do not hatch. If the eggs were destroyed the geese could renest.

Let's assume that eggs have hatched and we have an abundance of adult or recent fledglings. How are they controlled? At the Lincoln, Nebraska airport (like most), a person is assigned full time to keep critters like geese or turkeys away from runways, and several methods are used including pyrotechnics. My class got to shoot a special shell out of a 12g shotgun that went out about 100 yards and exploded like fireworks. (My class reviews were never higher.).

Another idea for reducing crop damage is to scare the geese away. Yes, scare them. Imo that works for a short time, but if we're talking Canada geese, as Arnold says, I'll be back. Of course, it's hard to tell if geese returning to a field are newbies or ones that were previously scared away. That could be an interesting scientific question.

Johan Månsson and colleagues recently reported on a scientific study in which they documented the behaviors of GPS tagged geese that were scared off crops, either by humans walking towards them or by a drone. In south central Sweden Graylag (or greylag if you're from across the pond) geese arrive in March and leave in September, and forage in ag fields or wetlands. The local government spends from 80,000 to 140,000 euros to reduce goose damage to wheat and barley, particularly.

Geese were captured, some when molting and thereby flightless, and fitted with neck collars containing solar powered GPS trackers. Humorously, at least to me, they conducted 299 "scaring events" on geese feeding in agricultural fields. They selected a collared goose to follow that had not been scared off in the prior 5 days, to reduce chances of it being forewarned. Before the scaring event, they randomly chose drone or walking to initiate the scare.

They found that geese moved about 1000 m (a bit over half a mile) after being scared off the field. Within the first few hours geese avoided the scaring site, but this aversion waned within 24 hrs (think trying to keep a teenage boy away from the fridge before dinner). After being scared from an ag field, geese tended to move to wetlands, but this effect was also relatively short lived. The authors pointed out that wetlands provide greater protection (i.e., they're a natural habitat), whereas greater food return comes from ag fields, hence the pressure to return. In addition, the authors found that although scared geese didn't return right away, other local geese recognized an open opportunity and moved into the "scare site." Interestingly, whether the scaring was done by a person walking in the field or a drone didn't yield different results.

Most scientific experiments have a control. In this study, an important aspect would have been to compare the behavior of scared geese that were not GPS collared, as possibly the handling of the bird and presence of the collar might influence subsequent scaring behavior; this of course is difficult if not impossible. Nonetheless the researchers concluded that the behavior of collared birds was realistic, even in lieu of a control group.

At this point, North American readers are shaking their heads and asking, why? Why not just have a liberal hunting season? This would redefine scared as dead (without the possibility of return) not to mention flock size reduction. I wonder if a bird in a flock from which comrades were shot would stay away longer, because they would have witnessed firsthand a tangible threat. The conservation status of the graylag goose is listed as "least concern" by the International Union for Conservation of Nature (IUCN). The authors mentioned that there is some goose hunting in this part of Sweden, but apparently not enough so as to explore other options like scaring.

The author's concluded that "coordinated and repeated scaring is necessary to deter geese from specific fields and, more broadly, agricultural land" and furthermore that "our study shows a limited efficiency of scaring by drone." So, cost effectiveness and time are issues. FYI, if you want to help reduce goose numbers and add decent table fare in Sweden, you need to pay a fee and pass a theoretical and shotgun practical exam and pay for a license for whatever guns you use. However, the bag limits are generous because there appears to be no bag limits. Hence your harvest depends only on your shooting prowess. Still, it's more involved to hunt geese in Sweden than in the U.S.

I know for a fact that walking into a field of snow geese in spring results in their immediate departure, for good reason. Snow geese have become quite wary since spring hunts were introduced for population management. It occurs to me that it might be a lot easier to scare off geese that have been hunted. Nonetheless, I found that these results provide an interesting glimpse into the behavior of geese that have been scared away, because without being able to identify individuals (unless you're another goose), you have no way of knowing who's who, or what geese actually do when scared off a field. Their study put a name, so to speak, to individual geese that allows learning their behavior and how they react to potential threats.

39. On A Cold Winter Day: Look For Snowy Owls

Cabin Talk.—If you're at the cabin in winter, and want to see something cool, look up local birding "hotlines" on line, and see if anyone has reported a Snowy

Owl. I remember clearly my first encounter with a Snowy Owl. I was driving in western Minnesota in Pope County, looking for winter birds, when I noticed a white object sitting on top of a light pole. My heart raced as I got closer and realized it WAS a Snowy Owl. It was obliging and stayed on the pole while I approached and took pictures of it, which I still have (somewhere, in a box,…). They are pretty tame, because where they spend most of the year, they don't see many people. Trying to figure out whether the owls are declining is a complicated task because of the biology of this species.

A popular place to find wintering Snowy Owls is at the Duluth, Minnesota harbor, where they perch on structures and watch for rodents hanging around the grain

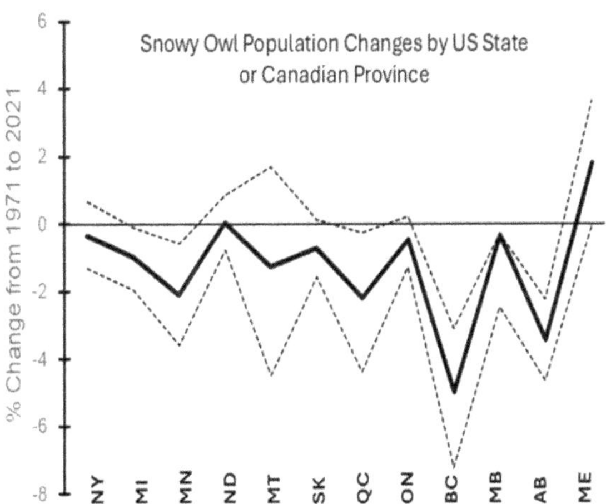

Graph 1. Dashed lines show the statistical upper and lower estimates of trends in population size of Snowy Owls

terminals. In "good" years, they are easy to find and in irruption years they have even ranged south to Arkansas, Oklahoma, and Texas. One was photographed in Bermuda in November 2013. This species is one of the diurnal owls, which you could guess because it breeds in the high arctic tundra where in summers it's light most of the day. Their distribution is circumpolar, with breeding occurring in the northern reaches of Scandinavia, Russia, Canada, and parts of Greenland. Snowy Owls undergo what birders term winter irruptions, when large numbers sometimes appear well south of their breeding range. The cause of these irruptions is not certain, but possibilities include arctic food supply crash, harsh winter conditions preventing them getting to lemmings and voles, or their population itself has erupted forcing birds south in winter because there's not enough food far north for all of them.

Because the irruptions are periodic and unpredictable, mostly no one panics during a winter when they are hard to find. So, how would you even know how their populations are doing? For most birds, ornithologists go out and count them, and once you've seen a Snowy Owl, they're easy to spot, unless they're sitting on a snow bank in a field. If the bird is an adult male, they tend to be almost pure white making they tougher to spot, whereas females and young birds are often pretty heavily mottled with brown. According to A. Hertzel, most birds seen during irruptions are the dark brown mottled first year birds, which makes sense as they would not be as experienced at finding food in the dark high arctic winter. Nonetheless, it would seem to be relatively straightforward to go out and count the number of Snowy Owl sitting out in the open.

Except that as I mentioned earlier, their nesting grounds are far northerly tundra, where there are no roads and it's difficult to census them on their breeding grounds. Another confounding aspect in trying to census Snowy Owls is that the breeding grounds reflect differences in annual food supply. Some radio tracked birds moved over 500 miles from their previous nest looking for a new spot to nest. That is, if you census the same area in consecutive years, are the numbers lower or did they move a few hundred miles to track lemmings?

A 2023 study published in the journal Ibis claims that we have overestimated the number of Snowy Owls worldwide by a hefty figure. The conclusion is based on some studies of breeding populations in Alaska, Canada, Greenland, Fennoscandia and Russia. Whereas it was previously thought that there were almost 300,000 individuals worldwide, this new study claims that the appropriate estimate is between 14,000 and 28,000. That's a wide range, but still a big discrepancy.

I looked up data on Snowy Owls from the Christmas Bird Count program for several northern states and some Canadian provinces (see graph) between 1971 and 2021. This program is run by the Audubon Society under the Christmas Bird Count label. It's not very scientific, but it provides a decent relative look at how populations are doing. If Snowy Owl population size was on average not changing, the lines for the upper and lower bounds and

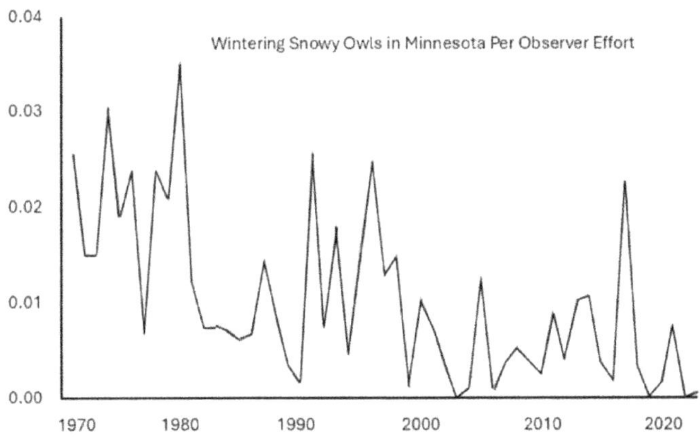

the mean should oscillate around zero (the solid line at 0). However, it is apparent that over the period of 1971 to 2021, most years show declining population size (i.e., the lines are below zero).

At a local level, the Minnesota Ornithologists' Union provides data on population trends in Minnesota. This chart shows the high variation in owl numbers seen each year, and if we focus on the same period as the other chart, there is an overall decline. Thus, several sources of information seem to be telling us that Snowy Owl numbers are declining. Whether this is a problem depends on several factors. All populations fluctuate, some go into decline and then recover and some do not; given that 99% of all species that ever existed are extinct, obviously the latter is quite common. We just don't want to be the ones who caused an extinction, as we've already done some of that (think Passenger Pigeon).

What could account for the decline? The authors of the new study point out the obvious ones, climate change and its impacts on food supply, collisions with vehicles (which owls don't see in the arctic) and pollution. At the very least, the new study points out that greater monitoring should be focused on this classy owl, despite the difficulty in doing so. I hope others get the same thrill I saw when I first saw this really magnificent arctic visitor.

OUTSIDE OUR CABINS

40. The Atlatl And The Demise Of Some Really Big Animals 12,000 Years Ago

Cabin Talk.—If you've seen hunting videos either on an outdoors channel or online you're aware that we have weaponry that is more than sufficient for taking down "big game." You know that early humans hunted mammoths and mastodons and defended themselves against dire wolves and cave lions. How in the heck did they do that without a large caliber rifle! Seems dangerous to be out and about with just a sharp stick, fortunately that wasn't the case.

The animals you might have faced 12,000 years ago on a daily basis provided some formidable creatures. As noted above, these animals included: an armored armadillo like creature called a pampathere (500 lb), an armadillo with a tail spike called a glyptodont (1,800 lbs), an elephant-like beast called a gomphothere (10,000 lbs), horses, a tapir (700 lbs), ground sloths (up to 6,000 lbs), mastodon (12,000 lbs), mammoth (13,000 lbs), American zebra (700 lbs), llamas, giant bison (4,000 lb), an extinct pronghorn, giant beaver, bison, white-tailed deer and mule deer. Of these large-bodied herbivores, only the deer and bison remain. There was a lot of meat on the hoof, and a cadre of large predators to keep them in check.

Scientists oscillate between two explanations for the extinction of these large plant eaters (and their predators). One is the Pleistocene Overkill Hypothesis, the other climate change. Climate change could have played a role, as at that time we were a mere 10,000 years after the retreat of the last glacier (one of many glacial advances and retreats over the last 2,000,000 years – in case anyone thought that climate change is new).A third idea envisions a combination of these two ideas. In the Pleistocene Overkill Hypothesis, it is assumed that early inhabitants of North America roamed the countryside and fed off these large animals.

But surely, early humans didn't just eat large animals that died of old age. Someone commented to me that the notion of megaherbivores driven to extinction by hunting seemed unlikely and suggested "if I'm hunting with a rock-tipped stick, I'm going to focus on smaller critters and vegetables for my main diet!" This in spite of noting that he was aware of some large kills, such as the well-known bison kill sites ("bison jumps"), such as the one at Ulm Pishkun, a cliff in Montana that stretches for a mile and has a layer of bison bones nearly 13 feet deep at its base. It is one of the largest of 300 such bison kill sites in Montana. These kills, however, were not with the aid of weapons. How could early humans take down a large animal, say a horse that once roamed North America?

At the museum where I work, the Nebraska State Museum at the University of Nebraska-Lincoln, we have an exhibit on hunting methods used by "prehistoric" peoples of the Great Plains. One of the main weapons was the atlatl (see figure), pronounced at-lat-tul. Atlatls were a ubiquitous part of the prehistoric hunting toolkit. This weapon, which can be described as a "launching stick" allowed a person to throw a sharp broadhead-tipped spear at up to 80 mph and hit an animal with considerable force sufficient to penetrate skin and vital organs. Not at all like a "rock-tipped stick."

The engineering was decent. The shaft, probably made from river cane, had a pocket at the end to fit onto the "knock" or knob on the handle (see figure) and likely some sort of fletching. There was often a weight called a bannerstone attached under the middle of the atlatl handle to increase velocity and kinetic energy. This tool basically extended the hunter's arm length and acted as an enhanced lever to propel the shaft at far greater speeds and force than throwing an arrow. On a par with a ball thrower for dogs or a handheld clay pigeon thrower. The atlatl allowed early hunters to take down big game and protect themselves against marauding predators.

No one is positive when and where atlatls were first used but atlatl is an Aztec word, and they used them to penetrate the armor of Spanish invaders (effectively pinning them inside their armor, ouch). However, there are similar weapons invented on other continents, and it's thought they got to North America over the Bering land bridge.

It obviously took some skill to become accomplished with the atlatl, but, then again, after practicing 9 months with my first compound bow, I missed the first deer I shot at. And I wasn't trying to feed my family or ward off a hungry lion (yes there were lions in North America at that time). Like my shoulder problems that led me to a crossbow, remains of a person from over 42,000 years ago showed arthritis in his right elbow, sometimes referred to as "atlatl elbow," resulting from years of launching these shafts!

Did the atlatl improve hunting success? Yes, quite clearly. We know from archaeological records that animals killed with atlatl included mammoth, mastodon, elk, bison, reindeer, camels and deer. We don't know the density of the megaherbivores, or how many people existed 12,000 years ago in North America. It's not likely any of the really big herbivores were as common as deer are today. But, the atlatl might have been at the root of the Pleistocene Over-kill, if that actually happened.

One might wonder why the big herbivores (and their predators) went extinct but bison, deer, antelope and caribou remained. These animals, with large latitudinal ranges and high reproductive capacity, had over 12,000 years to "rebound" from the loss of the other megaherbi-

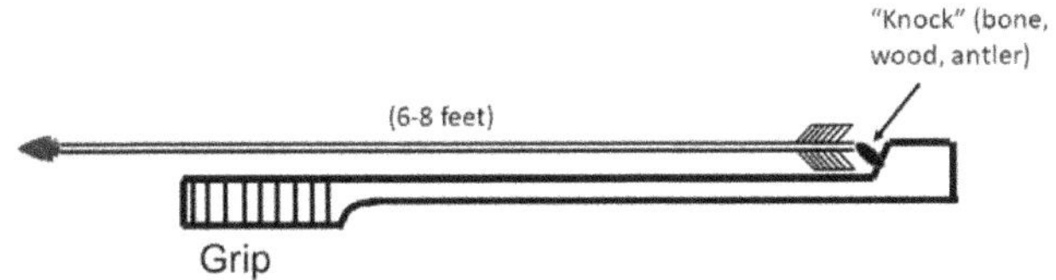

A diagram of an atlatl. The user would hop forward a couple of steps and launch the projectile towards his intended kill.

vores. Bison sustained great numbers even in the face of the bison jump sites. It wasn't until Europeans that their numbers plummeted.

Thus, no matter whether the really large megafauna were killed off by hunting, climate change or both (my guess), it is clear they had a weapon, the atlatl, to accomplish the former.

41. A Crisis In The Public's Lack Of Knowledge About Nature And What We Can Do

Cabin Talk.—Your neighbor asked you the other day what the blue bird with the crest was called. Shaking your head, you think about how the internet and its social media outlets has exposed a large fraction of the population that does not know much about the outdoors. As our population lives more and more in cities and gets outdoors less and less, knowledge of the identification, behavior and ecology of native species has eroded to the lowest point in our history. What to do?

Many of us who have spent time outdoors getting to learn things about nature are often dismayed by things some people say about animals and the things they do. On a naturalists site a few months back, there was an image of an osprey and an eastern kingbird in flight (see image). The author stated that this was an adult osprey taking its youngster out for a flight. Even if you didn't know it was

Osprey and adult Eastern Kingbird

an eastern kingbird, it's obvious it was chasing the osprey. To think this kingbird, coming in at 3% of the adult weight of an osprey was a young osprey, reveals a substantial disconnect with nature.

Another site mentioned finding dead beavers in which the feet and tails were cut off and possibly decapitated, and the author claimed it "seems as if they were killed for fun and tortured." How could you know someone tortured these beavers from the condition of the carcasses? A trapper discarded the skinned out carcasses, which then fed racoons, eagles, coyotes, foxes, etc. (this could be illegal on public land). Lack of knowledge of the outdoors lets the imagination run rampant.

A writer proclaimed that the beavers should be trapped and moved to a spot where they wouldn't interfere with human interests, assuming you could find a suitable place with landowner's consent. But, if we do this for beaver, it opens the floodgates. We'd need a tax increase to hire crews for relocating beavers, deer, red squirrels, coyotes and wolves, gophers and moles, bald-faced hornet nests, Canada geese (no one wants them anymore, we tried), Eurasian collared doves, to name some. I don't see a bond referendum passing to pay for these relocations. Sure people created human animal conflict, but relocation is not the answer – that ship has sailed. It would help if we had more undeveloped space, which ironically is what hunting organizations like pheasants forever and ducks unlimited provide.

A short video showed two ravens harassing a great horned owl during the day. One person commented "Not good. Owls need to sleep during the day so they can hunt at night" and "I feel so bad for it. It was truly annoyed by these two bullies." Where to start? The short answer is that ravens, crows, and owls have interacted for millennia, nothing important here. If you've spent more than a few minutes outside you've seen owls chased by all sorts of birds. Why? If a great horned owl found an unattended nest of a crow or another bird, it would without hesitation snatch a baby. One commented that it was too bad the owl didn't have a few friends to even the score. Oh my. Have you ever seen a flock of owls? They are fiercely territorial and this comment suggests a serious lack of outdoors experience.

A January post showed a squirrel with a naked tail, and someone claimed that squirrels chew off the hair to line their nests. How could anyone seriously believe this? Millions of squirrels breed annually, and if that were the reason for naked tails, we each would have seen it countless times. Almost no one has, and I'm sure I've seen thousands of winter squirrels. A moments thought should lead a viewer to the obvious conclusion that it is not adaptive to chew off your hair in the winter, and that there is something wrong with these squirrels! Unfortunately, there is a website run by a pest control company that claims that lactating female squirrels might chew off all of their hair to line their nests. This is bogus. Checking a few reputable websites, like by a DNR or extension unit, will point out that the cause of hair loss is a fungus or mites. A quick check of authoritative sources, like DNR web pages or university extension (e.g., https://www.purdue.edu/fnr/extension/hairless squirrels/; https://www.wildlifeillinois.org/sightings/squirrel with missing fur/), you'll find out that the cause of missing hair is either a fungus or follicle mites that cause Notoedric mange (which I learned from a professional mammologist who has worked in a famous museum for 45 years).

In the mammalian species account for gray squirrel (https://doi.org/10.2307/3504224), there is no mention of this behavior. If you look up what is in a squirrel nest you'll find these items: live green twigs, moss and damp leaves and an outer skeleton of twigs and vines – but not hair. Squirrels might use some molting hairs in their nest (and waterfowl routinely pull out down feathers for the nest, but they're not left naked). Looking at the images of squirrels lacking hair from large regions of the body, it should be obvious that intentionally stripping away large patches of fur is maladaptive, and instead is indicative a sickly animal (see the image in the Wildlifeillinois site). Squirrels do not pull out large patches of hair to line their nests. Always check sources, misinformation, even if cited on a web page that seems knowledgeable, is still misinformation, even if it's free.

An image was posted of a hognose snake eating a toad, with the response that nature "can be very cruel when some animals are eaten alive." Apparently, this person's wish is that any animal eaten by a predator had died of old age. It was pointed out to the author of the comment that it's been like that for millions of years, nothing cruel about it, it's the way it is. The response was "Nature is sometimes evil and the world sucks. Diseases kill people and that's part of nature. Would you say there's nothing wrong with that?" The response: "Diseases have been a part of life since its origin. To say that predation, parasitism, or disease is "evil" reveals a profound misunderstanding of the natural world. It's just as much a part of life as life itself. There is absolutely nothing odd or "wrong" about fatal diseases. The response (misspellings left intact): "If you truely belive that, you are disgusting." The distinction between nature and reality was lost.

What to do? These are common examples of the degree to which the public's knowledge about nature and the outdoors has eroded. As our population becomes more and more centered in cites, nature has become something to fear rather than respect and hold in awe. Some folks lack critical thinking skills (like the squirrel tail example above) and will dig in their heels if you tell them they don't understand the issue — after all they read two posts on FB and they regularly watch Disney type nature shows. Here again is the Dunning Kruger effect (people don't know enough to realize that they don't know what they're talking about). What though, if anything, has changed? In the past, social media didn't provide a platform for people to air their inaccurate, emotion-laden misconceptions about the natural world, and maybe we just didn't realize how far the general level of understanding of the outdoors had slipped.

For every story about a cute fawn or a finch eating from someone's hand, we need a corresponding story about a red squirrel taking each chickadee baby from its nest and eating it alive (at least initially). We need to be reminded that eaglets and baby herons sometimes kill each other, a phenomenon called siblicide. Hawks begin tearing a mouse apart before it's dead. Too often, people want to believe that the latter is an anomaly, but in reality, it happens millions of times each day, just as often as the warm and fuzzy stuff.

The public is shielded from the realities of nature. People who feed birds, probably unnecessarily because birds have lived for millennia without feeders, need to understand that each day raptors like the Cooper's hawk, which has now become fairly common in cities, eats on average two birds that frequent bird feeders. That's what hawks do. There are between 770,000 and 920,000 Cooper's Hawks (see Oddities chapter) in the U.S. and Canada, so if we go with 800,000 hawks, they kill 1.6 million birds a day. Gives a dual meaning to feeding the birds. But according to some, beautiful, majestic hawks are supposed to be on diets when they visit their yards. Ignorance of the laws of the land is not an excuse, neither is ignorance of the laws of nature.

It's past time to help people understand the realities of nature, irrespective of whether a person finds them pleasant or unpleasant. A first step would be a basic course in natural history that every school child should have to take. It would give equal time to predators, prey, diseases, and parasites and hosts. Nature is complex and no one person can know everything. Therefore, we have to teach people critical thinking skills. Before accepting something you read or hear, ask yourself "Do I know enough about this topic to have an informed opinion? Are the sources cited credible? Don't repeat something you yourself cannot back up because the longer the chain gets, the more people will believe it. Misinformation is the offspring of social media, and it's killing our understanding of the real world. On the other hand, I learn a lot from credible online sources. Teaching people the difference should be a goal.

42. Mosquitoes, Even Nature Lovers Swat Em, But They're Actually Pretty Amazing Animals

Cabin Talk.—Gazing at the winterscape outside the cabin as daylight finally emerges, a cup of coffee in hand, I stare at a tree rising up from the snow. I envision a morning 6 months ago, or one 4 months from now, and marvel how much things will change, and of all the living things that will be around and on that tree. Despite the winter serenity and seemingly simpleness, I am happy that it's a time with no mosquitos. Then my mind wanders, what the heck happens to them in the long Northwoods winter? Or, horror of horrors, what if they were active all year?

For now anyway. During the depths of winter, it's hard to imagine that they ever existed, or that they will miraculously appear next spring – shouldn't they all be frozen solid, dead, gone forever? Nope, I know they'll be back. Even if we thought there was a golden lining in really cold winters riding the world of mosquitos, I learned that cold doesn't affect the mosquito overwinter survival as much as lack of water next spring.

You've seen one, you've seen them all? Not really, a mosquito is not a mosquito is not a mosquito. Over 3,000 species have been described worldwide. They vary in size, patterning, and when they emerge. Some common ones are the Culex mosquitoes (several species), the summer floodwater mosquito, the Cattail mosquito, and the tree hole mosquito. Most people recognize a mosquito irrespective of what species it is, but fewer recognize the larvae (see photo), which can survive in water in old tires, trash, hubcaps, etc.

Surviving the Midwestern winter is no small feat, especially for a tiny insect. Most mosquitoes overwinter as eggs in the soil, and if you want even more bad news about them, buried mosquito eggs can remain viable for up to seven years. With warm temperatures and water, the eggs will hatch, our hatred rekindles, and our blood levels drop. Some species can hibernate overwinter as adults, even in cabins! Others can overwinter in water as larvae in a suspended animation, waiting for the water to warm. Typically males don't overwinter, just females. FYI, males have "fuzzy antennae" that make them stand apart from bloodsucking females – save your energy for swatting the females.

We've all experienced summertime outdoor gatherings and noticed that some people are magnets for the little demons, others not so much. Usually the former types are constantly shifting their seats at the campfire to be in the smoke plume.

What causes some people to be more attractive to mosquitoes than others? First some biology. Mosquitoes, as well as fleas, bedbugs, blackflies, ticks, horseflies, and biting midges, require protein to develop eggs, and blood is a good source of this protein. A female of some species can lay eggs every three days! Most male mosquitoes, however, feed on plant nectar. This might mean that the mosquito swarms we see are just half the number of bothersome insects!

Because it's do or die for a female mosquito to obtain a blood meal, there's been a lot of evolution of the behaviors, physiology and equipment they need. Overall, the task is to find a blood source, get a full meal and escape before being squashed. Judging by the number of live blood-filled mosquitoes I find in our lake place, they're good at it. Here are some things that make a mosquito an efficient parasite.

To find a potential meal, mosquitoes can sense CO_2, which is given off in human breath and through our skin, using receptor cells on the antennae and legs. Some people with a genetically elevated metabolic rate, those that are out jogging, or those that are consuming alcohol, give off more CO_2. They are likely the ones downwind of the campfire smoke.

However, animals aren't the only source of CO_2, and our female mosquito has to figure out how to distinguish car exhaust from your arm. They use other cues like the presence of lactic acid, ammonia and fatty acids to complement their CO_2 detection ability.

Mosquitoes also use vision to identify a host. By flying close to the ground, they can see hosts against the

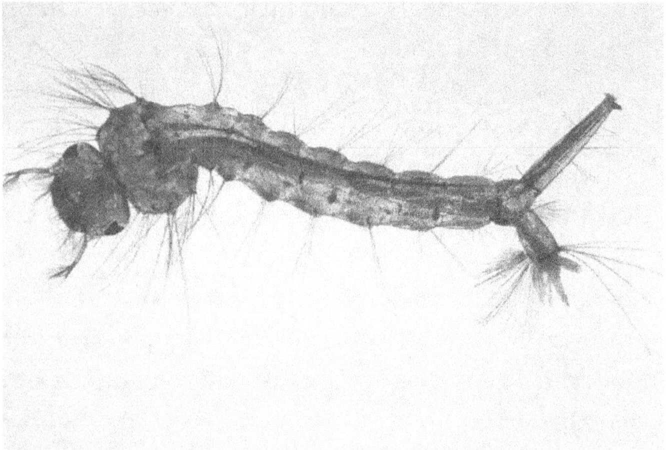

Larval mosquito. They feed on microplankton, algae and fungi in the water and some feed on other mosquito larvae! They stick the tube at the right out of water to breath. Removing water sources can keep numbers down.

horizon. Tip, wearing light colored clothes can help thwart the little buggers. They also can detect motion, so keep still, which is hard when they find you.

The needle, technically called a proboscis, actually is 6 needles. Two have tiny teeth at the tip to saw through the skin, two hold the skin apart while she inserts into the hole, and the main needle looks like the tip of a hypodermic needle, punctures your skin and finds a nearby blood vessel to puncture, and becomes a straw. There are special chemical receptors on the needles that sense natural secretions from our blood vessels, helping to hone the search for gold. The sixth needle secretes a substance to make the blood flow more easily (and causes welts). As she draws up blood, her body separates water from the blood, and excretes it out the end of her abdomen, making more room for red blood cells, her ultimate prize.

As a "parting gift," if the mosquito is carrying a virus or a parasite, it will be left behind. Most of them don't harm the insect but simply use the mosquito as a "vector" or traveling companion. For example, West Nile Virus, and equine encephalitis are spread by a species known as *Culex pipiens*.

Besides the many products that supposedly zap or repeal mosquitoes, what are their natural predators? Bats, dragonflies, birds, fish and turtles all snack on them. During the height of one mosquito hatch, I noticed very few on my morning walk, although there was a big swarm of dragon flies. Unfortunately, they didn't hang around too long and mosquitos reappeared.

My questions: I usually notice mosquitoes from their buzzing sound. You'd think that some mosquitoes would have evolved silence. Hard to sneak up on someone when you're buzzing away your presence. And what is the fate of the average mosquito? How many are going to find a blood meal and how many perish unfulfilled (pun intended)? How any vertebrate can escape undrained is a mystery. It's not uncommon in northern MN at the height of the insect season to see deer standing in the middle of a paved road, trying to escape the many insects that want to suck their blood.

43. Global Warming, Fact Or Fiction?

Cabin Talk.—Not many days go by without hearing about some aspect of global warming. Melting icecaps, species extinctions, birds breeding earlier, you name it. Still, it's sometimes hard in our climate controlled lives to grasp the reality of this phenomenon. Most of the information comes from far-away places, and too often, we tend to think local and not global. But it's about global climate and small increments of temperature, and it's important.

Most of the things we who frequent Northwoods cabins do are rooted in the outdoors. Whether it's hunting big game, upland birds, squirrels and rabbits, trapping furbearers, or chasing walleyes, panfish, muskies or bass, looking many of the most important things in our lives we do outdoors. And certainly those that don't hunt or fish but like to be outdoors listing plants or butterflies, making bird lists, hiking or skiing, watching all things nature have the same goals. A healthy outdoors that we can enjoy now and pass on to our kids.

You'd have to be living under a rock to be unaware that some people think there is a controversy about global warming. Unfortunately, vague statements, even from scientists, contribute to the perception of controversy. Let's try and be less vague.

When we speak of global climate change, we could divide it into four questions. 1) Is the earth warming? 2) If so, how fast is it warming and is it faster than ever before? 3) why is the earth warming and are human activities involved? and 4) does it matter? Let's go through them one by one and see whether failure to consider each in turn leads to overall misconceptions about global warming.

The fact that a mile thick glacier covered much of northern North America 20,000 years ago and is no longer there is pretty good evidence of warming climate. More recently, the earth has warmed in the last 200 years. This is no hoax. $97^+\%$ of all scientists agree. The earth warms and cools repeatedly on scales from a year to millions of years. Heck, the temperature warms throughout most days! But seriously, the images of the loss of glaciers in the Antarctic and reduced extent of the ice pack in the Arctic are real.

Some might quibble that it isn't about the "little brief fluctuations", when 20,000 years is exactly a "brief fluctuation" in geological time. Look at the graph I provided and tell me what the line looks like for the last

Pika. Image: makieni777 Pixabay.

450,000 years. There are numerous periods of rapid rises in temperature, and the most recent upward swing reached a lower temperature than the previous 4. Recognizing that there is lots of slop in the older measurements ("slop" or measurement error increasing with age of the record), we can for example see an increase of about 13°C between 340,000 and 332,000 years before present, but we don't know whether that change was concentrated or linear over that time interval, because we don't have readings every year. If the former, it might be in fact as rapid as today. We can project, however, that if the current increase of ca. 0.85°C per century continues, temperature would increase 68°C over the next 8,000 years, not 13 degrees. which would fry humanity if it remains. It would certainly provide important motivation for the air conditioning industry. I will say that climate scientists believe that the current warming trend is faster than ever before.

Does it matter that the earth is warming? Depends on your viewpoint I suppose. Perhaps a better question is whether the warming trend is having an effect on plants and animals. Here, the answer is absolutely yes. Plants are flowering earlier, birds are returning to nest earlier. To the extent possible, organisms will adapt to new climates. However, they're used to doing this over long periods, and if we're making the climate change at unprecedented rates, it stands to reason that many species will be unable to adapt.

A species highly vulnerable to our warming climate is the pika, a rabbit-like animal that lives in mountains of western North America. It doesn't do well when the temperature reaches 80°F, which is happening far more regularly than it used to. Biologists have documented that pikas are moving upslope to parts of the mountains that are cooler. Pikas have been lost from several places in the Great Basin, Sierra Nevada, and southern Utah, where there is no suitable habitat. They are considered indicator species of climate change. It would be a shame to lose them!

So for the sake of credibility, I prefer to say that 1) the earth is warming, absolutely no question, and it has warmed before, 2) it is likely that the earth is warming at an unprecedented rate, although it is hard to know for sure, 3) it's likely that humans are at least a part of the cause, and therefore, we should do what we think will help mitigate the rise in temperature, and 4) there numerous examples of effects on plants and animals from the current warming trend.

44. Oddities In The Field And Home

Cabin Talk.—I describe here two events that epitomize my old friend Larry's comment "The odd thing about rare events is that they sometimes happen." Put another way: "Someone will win the lottery, but it won't be you or me." There's also a lesson about Darwin natural selection. Sometimes things happen to animals that irrespective of how well adapted they are, kills them.

A past spring my wife returned from walking our drahthaars and told me that there was a hawk on the ground near a stream bank, flapping its wings. Being an ornithologist, I told her that the hawk had caught something and was keeping its balance by flapping its wings, waiting for its talons to cause the prey to expire.

Still, I was intrigued enough to head out and find the hawk to see if it was still there. It was, and it was flapping its wings when it saw me, just like she said. Ok, now it seemed odd. As I approached the bird, I saw something amazing. I realized that a Cooper's Hawk had attempted to make a kill of something that was in a small burrow. Unfortunately for the bird, it didn't see a 5 ft length of rebar that was sticking up out of the ground right next to the hole. The bird impaled itself and slid down to the ground, where it was trapped, but still alive.

I had not been to that spot on our property and had no idea there was rebar sticking up from the ground. Obviously, neither did the hawk. Upon closer inspection, I realized the bird had no chance to survive even if I managed to pull it off the rebar. Because I have State and Federal permits, I humanely euthanized the hawk (with a shotgun). After removing the rebar, I brought the bird to my museum (University of Nebraska State Museum), where I prepared it as a scientific study skin, and it is now part of the permanent research collection.

Reconsidering my friend Larry's comments, we probably observed the only Cooper's Hawk ever to skewer itself on a piece of rebar. Yes, Larry, rare events sometimes happen. And, we can also infer that nothing in the

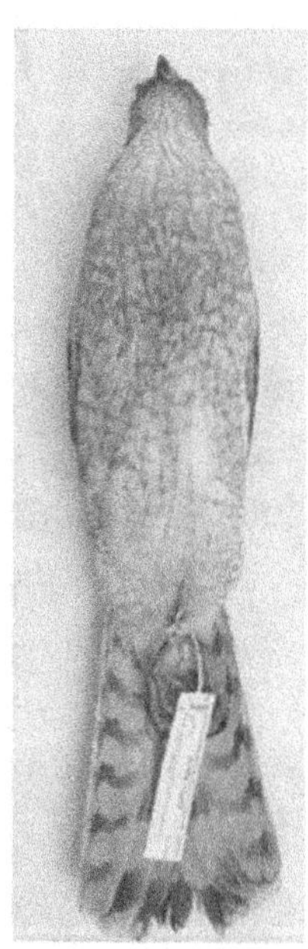

Cooper's hawk impaled on rebar, and the bird prepared as a scientific study skin by the author. It was an adult female and had recently eaten a common grackle.

evolutionary history of the remarkably well-adapted Cooper's Hawk prepared it for an unexpected piece of rebar.

Cooper's Hawks are often a source of dismay to people feeding birds because on average, each hawk eats about two nuthatch sized birds a day. Gives bird feeding a new meaning. There are estimated to be 840,000 of these hawks in North America, so they take over a million and a half birds each day. I'm all in favor of feeding the Cooper's Hawks, and for the record, our winter birds have lived for millions of years without bird feeders. But put out your feeders for the hawks.

My second observation strays into the bizarre. I visited our house on a northern Minnesota lake that we

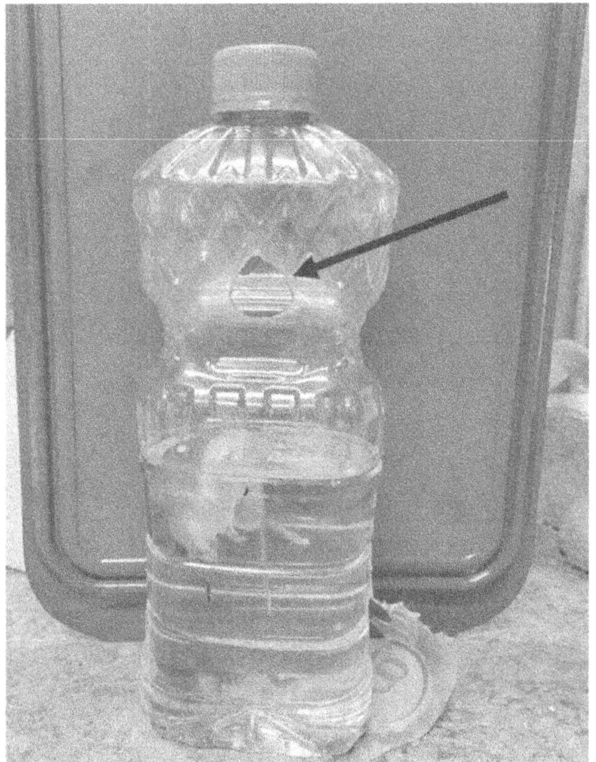

Deer mouse (genus *Peromyscus*) in a bottle. It got in but couldn't get out. Nothing in its evolutionary history prepared it for this challenge.

rarely visit in winter. I am used to finding signs of mice in the spring when I come up for the summer. One January,

I decided to make a short ice fishing visit and stayed at our place. I found an extremely high level of deer mouse (the only species I've trapped) activity, and they had gotten into cabinets they hadn't visited before. They chewed up a lot of stuff. They left droppings in the bathroom sinks, counter tops, bowls, and basically everything had to be cleaned.

After a couple evenings, we had caught enough perch to have a fish fry. I went to the closet where I keep vegetable oil and pulled out a half full bottle and started towards the stove. I was stopped in my tracks with what I saw. I called out to my friend "Come here, you ain't going to fricking believe this."

Floating inside the bottle of vegetable oil was a dead deer, or white-footed mouse. I quickly realized how it gained entry when I noticed that it had chewed a fairly large hole near the top of the bottle. I'm guessing it do so from the top of the bottle. It was floating in the oil (fact of the day: mice are buoyant in vegetable oil). I don't know whether it drowned in the oil or perished from exhaustion. In theory, it could have sipped oil for a while, since it apparently wanted to eat some in the first place. Imagine what it would have looked like in spring had that been true, an obese bloated deer mouse in a bottle.

We took this odd natural history observation as an excuse for another Manhattan and started to ponder possible outcomes of this unfortunate situation. What if the hole was a bit farther down and it could have pulled itself out? How popular would a vegetable oil saturated mouse be with his friends? We figured every one of his buddies would want a lick – they could take him on picnics for food and entertainment. He wouldn't need to leave a trail of breadcrumbs - he could just follow the oil slick. Maybe he could write a book – My (short) Life in a Bottle (of oil).

Fortunately, for me, it didn't chew below the oil level as then my cabinet would have been a real mess.

I'm sure that members of PETM (People for the Ethical Treatment of Mice) will find nothing funny about this. Scientifically, I think I can say that millions of years of evolution did not prepare this extremely successful species for this contingency. A colleague commented that this mouse was a top contender for a "Darwin Award" for rodents. One person asked if I had cut the hole on purpose to trap mice. What about mouse flavored oil for cats? Another friend commented "he won't make that mistake again."

Sometimes rare events happen.

45. Do We Need Lawns At Cabins?

Cabin talk.—I find it odd that sportsmen and women want clean lakes and rivers, lots of healthy fish, woods with deer and ruffed grouse, and open fields with pheasants. Yet some of these same people maintain "well-manicured" lawns by using dwindling water supplies, pesticides and herbicides, all of which are bad for the environment. I'm not the only one who sees the conflict here.

The U.S. has about 60,000 square miles of lawn. Therefore, turfgrass is the most widespread plant "crop" under irrigation in the U.S. In fact, estimates suggest that there are three times more lawn acres than corn acres. And "irrigation" here is key. I wondered how much water is used to keep lawns green. According to the EPA, landscape irrigation is thought to total 9 BILLION gallons of water per day. Any wonder why some aquifers and local lakes dry up?

So why are lawns less ecologically useful than native ecosystems? It's in the biology of their roots. Native prairie plants have extensive root systems, whereas Kentucky blue grass does not. The large root systems of prairie plants break up compacted soil, and when they die, they add organic matter and decompose into more carbon. Roots store a large part of the earth's carbon, and they clean our water as it filters from the sky through the plant roots on the way to the water table. Water quality follows carbon levels in the soil.

The psychology of lawns is also intriguing to me. When I was a kid, my mom and I had a nice dandelion crop, which spread to my neighbor's yard. He dug them from his yard and deposited them in a pile on our front step. For a quiet, polite and reserved man, this was a pretty aggressive move. I was unaware that dandelions could evoke such emotions, and my mother was embarrassed. Being partly red-green color blind, my favorite plants are dandelion and creeping Charlie, both of which sport colors I can see.

I agree that having a lot of tall dry grass next to my house is a good way to encourage fires, rodents, and I actually think it's good to keep the vegetation short near buildings. But the stats regarding lawns are depressing. Americans use over 30,000 tons of pesticides on their lawns. Of 30 commonly used lawn pesticides, over 15 are linked with cancer and birth defects, and are routinely detected in ground water. The National Cancer Institute says that children in households with pesticide treated lawns have a higher risk of developing leukemia (granted, probably a low level). Children are more likely to ingest pesticides during the first five years of life, and the effects can be seen up to 20 years later. Think kids crawling on the lawn during a picnic.

I recognize that a large number of people and businesses make their living because of our obsession with lawns. Fancy riding lawn mowers, and a host of other implements, make for big business. Gas, oil and equipment repairs are significant expenditures and support a large segment of our economy. Ever tried to find someone to repair your lawnmower during the summer? Seed, fertilizer, pesticides and herbicides all add up. We have college programs in turf management. I realize that lawn care companies and the turf industry provide lots of jobs and economic stimulus. But does this justify lawns, at least as currently kept? Not to me without significant changes.

I like my neighbors for who they are, not for their lawns. We need to reverse the attitude that a great looking lawn is a symbol of your value as a neighbor. We run the risk of running out of water while we poison what's left. I recommend looking at the University of Minnesota's website or others on low input lawns. Wouldn't you like to mow less? Here's what they say: "…University of Minnesota Extension recommends fine fescues, which are considered low-input species due to their drought, shade, and salt tolerance, as well as their lower maintenance needs (less mowing, watering, and fertilizing)."

There are many things to do to reduce the impacts on your lawn. Cut back on fertilizers and other chemicals; they find their way into the lake. If you convert your lawn to a native cover, it will aid native species. In a place we lived outside of the Twin Cities, with a pesticide- and herbicide-free "lawn," I would notice each spring that all five species of migrant thrushes were foraging in my lawn, whereas just across the dirt road, the chemically treated lawn (and all the posted warnings) had none. Isn't nature telling us something?

The last thing that should concern us is what our neighbors think, especially at the lake. Somehow, we've convinced ourselves that an unnatural, well-manicured lawn enhances our property. We need to get over that misconception. Just ask the thrushes.

46. Langmuir And His Currents

Cabin Talk.—I'd seen those parallel white lines on the lake's surface for many years and convinced myself they were the remnants of wakes that boats made across the lake. Somehow, they must line up. I tried not to think too hard because it was obviously not the case – there had not been a hundred or more boats traveling the same distance apart going in the same direction. When my kid asked me what they were, I tried to move the conversation elsewhere. Later in life, the same kid was now a PhD in fisheries management and was happy to inform me of the error of my thinking. "They are," he said, "Langmuir currents." The frothy white lines from Leech Lake, in this case (see photo), are the results of Langmuir currents, named for Irving Langmuir, who discovered similar phenomena consisting of windrows of seaweed in the Sargasso Sea in 1927. I could have looked it up and been educated but do we have to challenge everything we think we know? Ya, probably.

I always wanted to be smart in math, and I was until my 4th semester calculus at the university, when I was relieved to earn a C , which was the beginning of my biology training. Here then, is the equation that is called the "Langmuir number" that you can get by solving the equation below. So what, you ask? The equation explains the white lines on the surface of a large body of water. The actual mechanism is the playground of hard core math types, but I'll try my own explanation. In the northern

$$La = \sqrt{\frac{\nu_T^3 k^6}{\sigma a^2 u_*^2 k^4}}$$

hemisphere, the white traces are 0 to 20° to the right of the wind direction, and the bands are a helix of divergence and convergence at the surface. Doesn't make a lot of sense to me either, and don't ask what the variables are..

The Langmuir circulations are counter rotating cylindrical roll vortices. Not sure that helps either. They are transient, and their strength and direction depend on properties of wind and waves. In the cross sectional drawing, one can see the adjacent counter current rotations, and between these are the Langmuir circulations, or in our case, the frothy white lines or windrows across the surface of Leech Lake (or your favorite body of water). They are a natural consequence of certain conditions of wind and waves, which explains why they are not always present.

Naturally, one would wonder if they have any biological significance. In the oceans, they promote mixing of nutrients, and some marine microorganisms such as plankton congregate or get caught up in the circulations and attract things that feed on them. Once we realized what these white lines were, we tried to concentrate our baits under them. Didn't work for us. In other lakes, however, the foam lines concentrate Daphnia, which are also called waterfleas. Daphnia are planktonic crustaceans and have wide flattened legs that they use to push water over their filtering structures where they eat algae or bacteria. They

Langmuir currents on Leech Lake, Aug. 25, 2021; photo by author.

range in size from 2 hundredths of an inch to a quarter of an inch. They are extremely important part of lake food chains, and they are eaten by insects and small fish. You might not see them, but they're an important part of the aquatic ecosystem. Interestingly Daphnia "migrate" up and down in the water column over the course of a day, presumably to escape predators. People spend entire careers studying these little beasts.

At least now, when you see the parallel white lines on the lake's surface, you can confidently say "They're not from passing boats, but I forgot exactly what they're about." Maybe after a few years, which it took me, you can re-

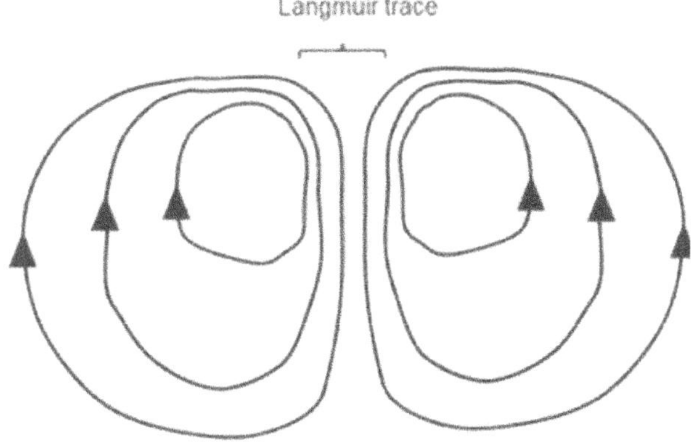

Cross sectional view of two opposing water vortices that cause the white streaks, or Langmuir currents, on an ocean

member Langmuir Currents. And, what the heck, try fishing along or under one, you might have better luck than we have had.

47. Neonicotinoids In Our Environment – Of Course, The News Isn't Good

Cabin Talk.—Even if I were only interested in hunting deer, grouse, ducks, pheasants, geese, woodcock, I should care about the environment at large because it affects all of these pursuits. Truth is, I am also an ornithologist, which I got into by becoming a bird watcher in 6th grade, owing to a remarkable teacher at Morris Park Elementary School in Minneapolis, John "Jack" Gilbertson, who took me under his wing. This fascination for birds he instilled in me led to a Ph.D. from UC Berkeley, a post-doctoral stint at the American Museum of Natural History, and professorships at LSU, UMN and UNL. Along the way, I learned about not only birds, but environmental issues.

What could be more important to us than our food supply? We have converted large stretches of native forests and especially Great Plains grasslands and prairies to agriculture and have successfully fed ourselves and a lot of the world. In striving to make agricultural practices efficient, we have faced challenges to make it safe. For example, DDT was used extensively in the 1940s and 1950s as an insecticide; the man who discovered its insecticidal use was awarded the Nobel Prize in 1948. Unfortunately, he did not know of its effects on unintended targets. DDT was banned in the US by 1972, owing to it causing many deaths in birds like eagles and peregrine falcons.

The DDT lesson drove a search for safer insecticides. What emerged was a class of chemicals called neonicotinoids, beginning about 1994. They work by getting from the seed coat into the plant's leaves, flowers, pollen and nectar, and their effects are long lasting. Their advertised advantage is that the dosage needed to control insect pests is considered harmless to unintended vertebrate targets, like mammals. They are now the most widely used pesticide worldwide.

More often than not, we don't learn from past mistakes. Among unintended targets are beneficial insects, like pollinators, who pick up the chemicals from nectar or pollen. Lest we forget, pollinators are how many plants have sex. I realize that corn is wind-pollinated, and soybeans are self-pollinated. But many plants harbor other native insects that do a service to the ecosystem and killing them upsets natural checks and balances. Some plant pests like spider mites and slugs do not die from ingesting neonicotinoids but when they are eaten by their predators, the predators die. This means that neonicotinoids can move through food chains.

The European Union banned neonicotinoids at the end of 2018 because of their threat to pollinators. Neonicotinoids are, however, widely used in North America and other non-European places. The Environmental Protection Agency (EPA) is proposing changes to their use in the U.S. What was once promising news about controlling insects didn't take into account the accumulation and persistence of neonicotinoids ("toxic load") in the environment, uptake by native plants, effects on unintended species, and their integration into food chains. And don't sell evolution short, bed bugs have evolved resistance to neonicotinoids.

Like DDT, unintended targets were identified relatively slowly. The early idea that vertebrates would not be affected is under scrutiny. Of course not all poisons are bad. We use low doses of Coumadin (rat poison) to help prevent blood clots in people, so it might be feasible that low doses of neonicotinoids could be fine. But many birds and mammals can get much higher doses from eating treated seeds.

A problem with widespread neonicotinoid use is that birds and mammals can find and eat spilled seeds. In agricultural practice, the neonicotinoids that coat seeds are dispersed to different parts of the plant, but if an animal eats entire seeds, they get the full dose, one that could reach acute or chronic doses (i.e., bad). In fact, eating a single treated corn kernel can kill a bird the size of a blue jay. These effects, however, vary from species to species. Lab studies have identified some sub-lethal effects. In birds, neonicotinoid poisoning includes wing drop, reduced hatching, low weight, and reproductive impairment. In mammals, there are premature deliveries, stillbirths and offspring deformities. Sub lethal isn't good.

Lab studies sounded the alarm about unintended consequences of neonicotinoid consumption in vertebrates, but what happens in the wild? Are neonicotinoids really a problem, relative to their value in controlling agricultural pests? Here's where some clever work by Charlotte Roy and Pamela Coy from Minnesota DNR in Grand Rapids, Julia Ponder from the U Minnesota College of Vet Medicine, and others, comes into play.

Their research team studied actual and simulated seed spills in agricultural areas of western Minnesota. They could identify treated seeds because by law they are col-

Corn treated with neonicotinoids You could see the colors if the image wasn't B&W!.

ored (pink, blue, green, purple) to prevent accidental feeding to livestock or people (that's telling right there). The team put down frames measuring a meter on a side at random spots in the middle and at the ends of field rows, where more seeds would be found as the planters make sharp turns. Over 100,000 acres of corn, soybeans, and wheat (each) were surveyed. Many spill sites were visible from roads, where hoppers were refilled.

To get some idea of seed longevity, hundreds of treated seeds were placed on fields and monitored for 0, 1, 2, 4, 8, 16, and 30 days, to judge the effects of levels of UV, microbes, rainfall etc. Trail cameras allowed the team to determine how long it took birds and mammals to find the spills. Average time to finding spills was 1.5 days for birds, 2 days for mammals.

Lots of animals liked spilled seeds: ring-necked pheasant, Canada geese, American crow, mourning dove, wild turkey, blue jay, brown thrasher, black-billed magpie, rose-breasted grosbeak, various sparrows, and white-tailed deer, black bear, raccoon, rodents and squirrels, Eastern cottontails, red fox, white-tailed jackrabbit, striped skunk and domestic cat. So, apart from cats, which is a good thing, lots of other animals were ingesting neonicotinoids.

The results were clear. Neonicotinoid treated seeds are common on the landscape, and wild birds and mammals find and eat them within days, while the chemical is still coating the seeds. The team directly observed hundreds of large spills, and never any attempts to clean them up, for at least five days. The authors stated "Our findings not only refute the idea that wild animals will not eat treated seeds, but unfortunately document that good seed stewardship practices were not always followed, despite clear warnings about dangers to wildlife on product labels."

In a follow up study, Roy and Coy attempted to answer the potentially more important question of whether the birds and mammals that came to the spills actually ate enough treated seeds to be harmed. They used different concentrations of neonicotinoids, and set trail cams to record video so they could count (or estimate) numbers of seeds eaten. From their 40,959 videos, they found that birds often didn't eat seeds the first day they found them, possibly because of the neonicotinoids, whereas some mammals started eating at first sight.

The results were good and bad. No animals consumed enough treated seed to reach lethal levels. However, 7 species of birds and 3 mammals ingest enough seeds to yield sublethal levels. Thus, these studies established key links between known sublethal effects in lab studies and documented consumption of treated seeds in the "wild." I do not know whether ingestion of neonicotinoids will be a major threat to wildlife health, but the threat to pollinators is very real. These findings coupled with exposure to unintended plants and insects makes the search for a safer insecticide pretty important.

The EPA is actively investigating the threats to unintended animals and has issued some guidelines. And some states have ruled that neonicotinoids cannot be used on lawns, and other non-agricultural plants.

48. Things That Happen With A Receding Hare Line

Cabin Talk.—I'm guessing that few people spend a lot of time thinking about rabbits. My experiences with rabbits are reasonably few and far between. Mostly I hear about them eating people's plants, and one winter I had a couple destroy the bark at the base of some apple trees I had planted. I had put sheet metal around the base but hadn't accounted for the snow level rising above them. I know a few rabbit hunters. A few years back our young drahthaar caught a large cottontail, and out of respect, we turned it into Hasenpfeffer. Culinary distractions aside, there are some interesting scientific things about rabbits and global climate change that might be unexpected.

21,000 years ago, there was a glacier as much as a mile thick that extended south to Nebraska, covering most of the Midwest. And there have been many such glaciers over the last 2 million years. Why they recede is clear, the earth goes through warming cycles. Hence, warming on a global scale is standard stuff, happens in a big way, and frequently. One clear consequence is that habitats shift as the climate warms. Conservationists recognize that areas set aside for a species today will likely not harbor the same habitat in 100 years, and in effect, be useless for the target species.

During the evolution of most animal species natural selection has built in some fudge factors that allow small adjustments to events in the annual cycle. We all know that some springs come earlier than others, and it would beneficial for birds to take advantage of a year in which nesting sites and insects for young are available earlier than most years. They do, unless it's way too early.

We think that the basic "clock setter" is photoperiod, that is, daylength. In experimental setups, you can divide birds from wintering flocks into separate groups, and by giving one group more "daylight" in captivity, bring them into breeding condition when the other (control) group is still in their winter phase. Daylength, unlike climate, has not changed over the last many years (although that is not true for the earth's magnetic field).

One documented effect of climate change, as opposed to subtle annual shifts, on animals was reported by Sean Sultaire and colleagues from UW Madison in 2016.

They found that snowshoe hares basically disappeared from the Sandhill State Wildlife Area in Wisconsin in the early 1990s. The reason seems simple enough, warmer winters means less snow in the southern part of the hare's range, particularly later in the winter. When a white hare is on a snowless background they're easy pickings for predators. Although the range has not receded northwards all that far, one can imagine what could happen over thousands of years. But we might say, hare today gone tomorrow, we have a receding hare line.

A paper appeared in 2022 authored by Evan Wilson and colleagues, also from UW Madison. They pointed out that snowshoe hares are food for coyotes, bobcats, fishers, and large owls, and in their absence, these predators might have switched to other prey, like porcupines and ruffed grouse. In the words of Wilson, there has been a ripple effect through the predator-prey community of the northward receding line of snowshoe hare distribution.

Or is this just speculation? Anyone who has trapsed about in Northwoods grouse country knows that populations fluctuate wildly, and in this hunter's opinion, the numbers have not been as high in Minnesota as they were in the mid-1990s. It is known that overwinter survival for ruffed grouse is tied to having enough snow for nighttime burrows. As for porcupines, I keep as far away as possible so that I'm not taking my drahthaars to a vet.

Being good scientists, Wilson and colleagues realized they needed some data. In the area without hares, they found that ruffed grouse populations dropped and that there were relatively few juvenile porcupines. The latter is explained because most predators avoid mature porcupines, instead preying on, wait for it…., porcupettes.

Being even better scientists, the researchers realized that they were still dealing with correlations and needed to do some experiments. Obviously, experiments are not trivial at the landscape scale of hares and their predator community. Still they persevered and captured about 100 hares from another location, fitted them with tracking devices, and introduced them into the hareless (I loved writing that) Sandhills area, and monitored what happened to grouse and porcupines. The idea is that predation on these two species would drop when predators had hares for lunch.

They captured 59 ruffed grouse (actually their colleague Amy Shipley did), determined age and sex, and fitted them with VHF radio transmitters that had "mortality sensors"; upon detection of a mortality event, they found the carcass and determined the predator, if possible. Porcupines were captured in traps, sedated, examined for sex and age, and injected with a PIT tag for individual identifi-

Photo: Evan C. Wilson

cation. Furthermore, in the spring when females give birth, they searched for the porcupettes and injected a PIT tag as well, and later recaptured them an added a VHF transmitter. Ready, set, go - what happened?

Of the 59 grouse, predators killed 32 and another 9 died from undetermined causes. There was not a strong relationship between predation on grouse and the presence or absence of hares, and in fact grouse tended to decrease over the course of the study, hares or not. This was attributed in part to the lack of snow, which grouse use to thermoregulate in winter, and the fact that grouse were mostly preyed on by avian predators. Also, grouse continued to cycle after hares disappeared from the study area, although the highs and lows were less than in prior years.

What about porcupines? Predation on porcupettes was the main cause of mortality when hares were absent, and when hares were introduced, predation on the porcupettes decreased by 60%. It was also suggested that when hare densities are low, predators like fishers switch to porcupines, which could actually buffer the few remaining hares. Nature has lots of connections.

What happened to the 100 reintroduced hares? It wasn't good for them, because by the end of the year (2017), there were just 20 left. Thus, this clever study went beyond the effects of climate change on hares and showed that the loss of hares had cascading effects on other species in the predator-prey community. I can't imagine the trials and tribulations these researchers went through to pull off a study of this magnitude. And a little compassion for the predators, who, still hungry, were heard to remark "Nope, there are no hares on my chinny chin chin."

49. Are Opossums A Tick's Worst Enemy?

Cabin Talk.—When you hear something often enough and from seemingly credible sources, your mind tends to file it away as fact. You come to its defense against anyone foolish enough to challenge what you now think to be a basic fact. I've reached the age at which when I hear something that contradicts my own belief on a subject, I now do a little fact checking just in case I was wrong all these years. What follows is one of those episodes.

Let's say your good friend, long time hunter, bird-watcher, all around naturalist tells you to leave opossums alone because they eat lots of ticks, like 5,000 or more each year. You kind of assume given their knowledge of the outdoors and animals large and small that they are simply stating factual knowledge, and you file it away in your memory banks. You were not tempted to fact check this statement.

Neither was I.

But the more I heard about all the ticks that opossums eat, the more I wondered how this possibly could be true. I mean, how could there be any ticks left in the world given the tick eating prowess of opossums and the large number of opossums? One would have to have a death wish to be reincarnated as a tick.

Opossums are synanthropic, meaning they thrive in a human dominated landscape. They are carriers of dog ticks, blacklegged tick, lone star tick, groundhog tick and rabbit tick. Yes, ticks like opossums, which you think would not be an adaptive behavior on the part of ticks given the common knowledge that opossums are veritable tick vacuums. Although opossums can carry diseases that affect people, like babesiosis or Rocky Mountain Spotted Fever, they are the "definitive host" for "equine protozoal myeloencephalitis" a fatal horse disease. Opossums are not appreciated by horse owners, although it is now thought that the threat to horses is not as great as once believed.

I knew there was a study by Richard Ostfeld of the Cary Institute for Ecosystem Studies in Millbrook, NY. Ostfeld, a former grad student colleague of mine, is an expert on infectious diseases like Lyme disease. He dosed white-footed mice, squirrels, chipmunks, veery (a thrush), catbirds and opossums with 100 ticks each, and sat back and watched what happened. Well, he came back later and counted ticks on the animals. Kind of like pulling the wings off flies, but with a scientific agenda.

His response pertaining to opossums was "I had no suspicion they'd be such efficient tick killing animals." During the process of self-grooming, opossums find a tick, remove it and eat it. Kind of like growing a garden on your own body, although at your body's expense – maybe recycling is a better description. By doing some simple arithmetic, Ostfeld and colleagues concluded that an opossum would consume 5,000 ticks in a tick season, the length of which obviously depends on where you live. If the tick season were 90 days, that would be over 50 ticks consumed each day by an opossum, which should be relatively easy to discover.

This analysis by Ostfeld's group is the source of the myth that opossums are like your robotic vacuum, going through the environment ridding it of ticks. This always made me wonder why on my property that has lots of opossums my dogs come back loaded with ticks. Is my property home to lazy opossums? And, how come I've never seen opossums doing closely spaced transects through my property, back and forth, sweeping the vegetation for ticks? If they are ridding the world of ticks, that's what it would take, imo. My sample size of 1 property doesn't support their tick-ridding abilities, but I recognize the limitations of n = 1.

What I envisioned is not actually what Ostfeld proposed, and to be fair, the popular press took his observations and quickly made it an urban legend; social media fanned the flame. A rallying cry, if you will, for leaving opossums alone. They are, after all, pretty amazing. Our only marsupial, rapidly colonizing northwards over the last century, and having mastered the art of playing dead. Unfortunately, for them, they look like giant white rats, and it seems that a relatively small proportion of the public likes having them around.

Obviously, I didn't muster up the time to go out and study opossums and what they eat. But someone else did. Cecilia Hennessy and Kaitlyn Hild published a scientific paper in the journal *Ticks and Tick borne Diseases* entitled "Are Virginia opossums really ecological traps for ticks? Groundtruthing laboratory observations." Finally, someone willing to test the oft cited lore about opossums and ticks.

Their methods were straightforward. It turns out that to determine the number of ticks eaten by opossums one searches through stomach contents (or feces). By looking through a microscope, one looks for tick body parts and then determines how many ticks were represented. This would not work for soft bodied things, like spiders, but ticks are pretty durable, as you remember from trying to crush them between your fingernails, so tick carcasses can be expected to persist in stomach contents.

Hennessy and Hild obtained stomachs of 32 opossums from central Illinois. And, drum roll, they found exactly zero ticks in their stomachs. Most of the opossums studied died during tick season and not mid-winter when you might expect few ticks. But what they found was that opossums like to eat beetles, nematodes (some of which were parasitic), small mammals, and roots, in support of the understanding that opossums are generalist scavengers/predators. Opossums do have a mouthful of teeth, which would make quick work of a rodent.

Image and caption from social media purporting to show a Virginia Opossum removing ticks from a white tailed deer, whereas the image shows nothing of the kind.

They didn't stop here, as they surveyed the literature and found 23 scientific papers on opossum diet, and found the same result, opossums are not eating many if any ticks.

The authors explored reasons why the Ostfeld study might have concluded that opossums were tick vacuums. One idea is that animals in captivity are bored because they are restricted from normal activities, and resort to doing things they might not do naturally, like eat the ticks off their bodies. Another idea is one based on the economics of foraging – it might take more energy to digest a tick with its hard exoskeleton than is returned from eating the tick.

The authors suggest a relatively surprising conclusion, that in fact, opossums might in fact preferentially *not* eat ticks!! How's that for a reverse urban legend?

Lastly, the authors noted that their study was motivated by the many memes in the popular press that touted the ability of opossums to remove ticks. One particular image they used was one I'd seen several times (see image). I was kind of jumping in my seat when I read their appraisal of the image: "viral meme that made the claim that the photograph shows an opossum benefitting a white-tailed deer (*Odocoileus virginianus*) by eating a tick off its face (Fig. 1b). There are no ticks visible in the photo. The claim is unsupportable…" I couldn't agree more.

But, if you don't question the image, it might make sense if you believe in the urban legend about opossums and ticks. The moral is pretty clear, question everything, fact check. So for now, when you hear someone say that opossums are tick destroyers, gently tell them it ain't so.

50. What Does Mammoth Taste Like?

Cabin Talk.—Ever dig through your freezer and find some unlabeled bag of meat at the bottom? What could it be, you wonder, and you go through a number of possibilities. Somone, hopefully not me, forgot to write on it, or more likely assumed it would be eaten in a few days, several years ago, so didn't bother to write on the bag. More likely, it's now trash or doggie appetizers.

A trait of museum curators is that we hardly ever throw anything away. I can find a potential future use for just about any worthless specimen, such as those in very poor condition or lacking any data about where, when and how they were obtained. Even perfectly good specimens can have unanticipated future uses. When the world noted the 100th anniversary of the passing of Martha, the last known passenger pigeon who resided in the Cincinnati Zoo, many articles and analyses were done. I was part of a group that obtained DNA from the toepads of dried museum specimens, and we were able to sequence a large part of the genome, finding that the population history of passenger pigeons was one of massive ups and downs. I doubt that anyone ever dreamed that this could be done when they collected and prepared the museum specimens in the 1870s.

In the collection of the Peabody Museum at Yale University is a chunk of meat that was from the main course served at an Explorer's Club dinner in New York City on January 13, 1951. This specimen, Yale #19475 has become somewhat infamous. The reason is that at the time, diners were told that they were eating the extinct giant ground sloth, *Megatherium*, that had been taken from the Alaskan permafrost on Akutan Island by polar explorers Father Bernard Rosecrans Hubbard and Captain George Francis Kosco (US Navy). Later, the Christian Science Monitor re-identified the main course as mammoth, which stuck in culinary lore. These two species did indeed occur in North America, although the sloth was not known from Alaska, but both species were part of the massive die off of large mammals (the "megafauna") 10,000 years ago.

Couldn't the story be credible? Specimens of mammoths are regularly found frozen in the permafrost, and many a tale has been told of the meat being thawed and eaten. Others, however, are more skeptical. In an early scientific account of the discovery of a well-preserved mammoth frozen in Siberia, the meat was described "as enticingly red and marbled but smelling so putrid that researchers could only tolerate a minute in its proximity." Well, at least the first part of the description holds some promise?

Could the Explorer's Club have found a specimen of either ground sloth or mammoth in Alaska that was still edible, and served it up at their 1951 dinner? Certainly, many people thought it was true, although accounts vary subsequently as to whether it was mammoth or giant ground sloth. Club members even talked of a second extinct-meat dinner. As an aside, apparently these dinners were not to be missed because not only did membership include Teddy Roosevelt and Neil Armstrong, but sometimes fried tarantulas or goat eyeballs were served as hors d'oeuvres.

Enter Paul Griswold Howes, a member who could not make the 1951 dinner. Howes was a curator at the Bruce Museum in Greenwich, CT, and he managed to obtain a sample of the main course, which he labeled mammoth. The specimen, preserved in alcohol, later made its way to the Peabody Museum.

Fast forward to the recent. Matt Davis, a graduate student at Yale, became fascinated with the specimen and its true identity. Eric Sargis and Jessica Glass joined the discussion, and Jessica realized that because ethanol is used as a DNA preservative, they might be able to recover

DNA from some of the less well cooked portions. The DNA would solve the debate and identify the animal that was the main course.

Their paper appeared in the scientific journal PlosOne. In short, they were able to extract some DNA from the meat, and they targeted a short piece of DNA from a commonly studied gene in the mitochondrion called Cytochrome *b*. A stretch of 300 base pairs was tar-

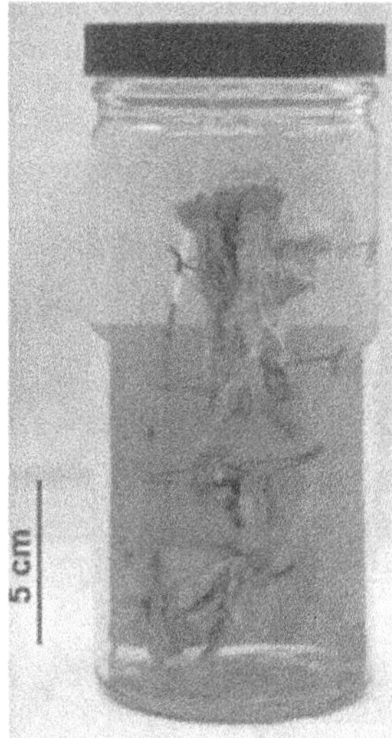
Actual specimen from Yale University that gained fame, erroneously, as mammoth that was purportedly served for dinner at the Explorer's Club

geted because it is an easily studied bar code that reveals the identity of most vertebrates, both those around today and those that have gone extinct. Thus, once you have the DNA sequence from the mystery meat sample, you find a match in the public data base (GenBank), and voila, you have your answer. They solved the question.

The answer was green sea turtle (*Chelonia mydas*), not mammoth or ground sloth. It turns out that the original perpetrator of the myth actually confessed, in a roundabout way, when he wrote that he may have discovered: "a potion by means of which he could change, say, *Cheylone mydas Cheuba* [sic] from the Indian Ocean into Giant Sloth from the 'Pit of Hades' in The Aleutians." In fact, because of this statement, Glass and colleagues targeted green sea turtle, so the search was short. Although there are genetic differences between populations of the turtle from the Pacific, and Atlantic, and Mediterranean oceans, the sequenced DNA was too short to identify the geographic origin of the dinner fare, although it was able to identify the sample 100% as green sea turtle. Today this turtle is endangered, but at the time it was eaten fairly commonly; there's probably a link.

It turns out that the club had also served turtle soup at the dinner, and it retrospect it is pretty obvious that the meat was turtle also. Glass and colleagues remarked: "Our archival research suggests that the prehistoric meat served at the 1951 ECAD was a jocular publicity stunt that mistakenly wound its way into fact." Surely this must be a unique instance? Actually, there was once a magazine account of the hunting of a live mammoth and its subsequent donation to the Smithsonian, which given strong public wishfulness, forced the museum to issue a public statement that the hunt never existed.

Still, there's some relief to the curators ethic of save, save, save. Glass and colleagues ended their paper by pointing out that Howes "probably never anticipated that one day the several nondescript pieces of meat he saved would finally lay to rest the myth of The Explorers Club "mammoth." As for my title, the jury is still out and may always be, until someone clones mammoths, which as you might know is underway.

51. Step On A Rattlesnake, See What Happens

Cabin Talk.—Going hunting or hiking in a state with venomous snakes? Ever wonder just what it's like to be bitten by a rattlesnake? I'm pretty sure that people with snake phobias simply won't go places where there are venomous snakes. Even where I live currently around Lincoln, Nebraska, there used to be native rattlers, but they live there no more, having succumbed to the hoe.

Snakes live a precarious life. My grandfather had a pathological fear of snakes and would dismember garter snakes in his yard with a shovel. He grew up in Ohio with rattlesnakes and the copperhead. So his fear was reinforced by actual exposure, which led my grandmother to tell me to leave him alone, it was his childhood (I have no idea whether grandpa had any actual encounters).

Some evidence suggests that we are born with a fear of snakes. Personally, I have gone both ways. I once kept snakes as a teen, which mostly ended when my large garter snake (note, they are not "gardner" snakes) escaped its cage and lived somewhere in our house for a couple of weeks. This sent my mother into a snake fearing frenzy in which every step she took was preceded by careful inspection of the area around her next step. I also kept a large water snake that liked to bite. Then I spent time in places with venomous snakes, rattlesnakes and cottonmouths in South Texas and Fer-de-Lance in Veracruz, Mexico. I would have some pretty disturbing dreams of stepping out of a truck at night and being surrounded by rattlesnakes.

Today, I won't kill a snake, unless it's venomous and a direct threat. But I confess, they're not high on my "most liked" list.

There is some reason for a fear of snakes, as the worldwide annual fatality toll from snakebite is at least 90,000. This doesn't include nonfatal bites that lead to 400,000 amputations and other health problems. In the US there are between 1,000 to 10,000 "envenomings" annually, but in Mexico it can be ten times that. Arizona records about 240 bites each year. However, the annual number of deaths from snakebite in the US is five.

In my admittedly nonexpert opinion, the innate fear of snakes could have at least two basic causes: snakes that are venomous and snakes that might eat you. There was a time during which humans ate snakes and snakes ate humans. These snakes include large constrictors like reticulated pythons. The incidence of human eating pythons today is very low but not zero. In 2018 an Indonesian woman of small stature was checking on her garden and attacked and eaten by a 21 foot python. Locals found the snake the next day, killed it and extracted her body. Gruesome for sure.

Just a BTW, the constrictors snakes kill mostly by suffocating the victim, not crushing it. Each time their prey exhales, they tighten their grip, which makes it harder and harder for the victim to inhale, killing by suffocation rather quickly.

You just tried this, didn't you?

A long-term battle between snakes and humans could have led to an evolutionary response in which we're born with an innate fear of snakes, before they could have learned to fear them from parents. This is the classic nature (genes) versus nurture (learning) argument for behaviors. Currently thinking is that human babies are curious about snakes but are not born with hard-wired fear. They learn it quickly from parents, who most likely have themselves learned to consider snakes as dangerous. Like my grandfather, who instilled in my dad the notion that the only good snake was a dead one.

The other reason for snake fearing is a venomous snake. Like most people my age, I grew up calling them poisonous snakes, which is wrong. I have eaten rattlesnake and it's not poisonous. Surely people have learned that venomous snakes are to be respected and avoided. There is a group of herpetologists (reptile biologists) who are fond of venomous species and often keep them as pets. One of my co-workers was the city certified snake remover in Baton Rouge, Louisiana, and he often kept cottonmouths in his bathtub. Unfortunately, when on an expedition to Myanmar in Southeast Asia he misidentified a deadly multi-banded krait, which bit him and he died 30 hours later.

When I was an undergraduate, I participated in a bird collecting trip to Veracruz, Mexico and my advisor told me before I left: "When you're walking in the forest and come to a log, don't step over it. Step on top and look down to see if there's a snake laying alongside you might get bitten by." There are multiple deaths annually in Mexico from this snake (see image of me holding one brought to us dead by locals in 1975). To this day 50 years later I don't step over logs even in the Northwoods.

What about my dreams and how likely is it you'll be bitten. A study by Cale Morris cast some perspective on bites from venomous snakes in the U.S. If you're like me, you assume that if you step on a rattlesnake, a cottonmouth, or a copperhead, it will bite you. That's why we wear chaps in snake country. Morris' fascinating study was focused on rattlesnakes, and some information was provided that probably conflicts with what we all think: rattlesnakes aren't aggressive, and they don't rattle before striking, instead the rattle is a warning. Only about 1% of ratt lesnake bites are fatal, and some bites are "dry bites" where no venom is injected. Baby rattlesnakes are venomous but not as venomous as in popular legend (Morris commented, would anyone say, "Watch out it's a puppy?"). Lastly, rattlesnakes are not losing their rattles because of people.

What Morris did was simple but ingenious. He concocted a fake leg with a boot and a covering portion of blue jean up the leg, all on a large pole. When he encountered a rattlesnake, he "stepped on it" with the fake leg and recorded the snakes reaction.

Out of 175 snakes that were stepped on with the fake booted leg, only 6 (3.4%) bit the leg. The other 169

Author, at age 22, holding a dead Fer-de-lance killed by local residents in 1975 in southern Veracruz, Mexico; specimen is in Bell Museum, University of Minnesota.

tried to get away, froze in place or wiggled. That's reasonably reassuring, but some people criticized the study by pointing out that the force generated by the fake leg/boot was not close to the weight of an adult stepping on the snake. Also, rattlesnakes are pit vipers and have heat sensing organs to detect prey, obviously the artificial leg lacked heat. You still might want to wear chaps but maybe my dreams can dial it down a notch.

If you are bitten, do not suck out the poison, ice it, or apply a tourniquet. Get medical attention asap. You can carry antivenom although there is not a one-size-fits-all antivenom, so you need the write mix for the snake that bit you. Obviously, you have to identify the species of snake.

Bottom line, snakes are a natural and important part of our ecosystems (except for example the pythons in Florida) and should be left alone. They are native predators just like hawks. We don't shoot hawks; we shouldn't kill snakes. Sorry grandpa.

52. PETA: Well Intentioned, Biologically Naive

Cabin Talk.—I admit to being a bit behind on my familiarity with PETA. The last thing I remember was that we all chuckled when we heard that in 1995 Michael Doughney registered the domain peta.org as "People Eating Tasty Animals". Alas, in 1996, Doughney was sued by the extremist organization People for the Ethical Treatment of Animals and ultimately was forced to surrender the domain.

Once you've heard that PETA stands for "people eating tasty animals," you can't unhear it, so there's that at least. It's still what I think of, almost 30 years later, when I hear or read PETA.

To be fair, there are examples of people doing inhumane things to wild (and pet) animals, and I think all hunters would stand with PETA against such behaviors. And, despite where this will go shortly there is one thing that PETA does that I HUGELY applaud.

And that involves their stance against the program called Trap Neuter Release (TNR) for unwanted cats. TNR is an ecological nightmare foisted upon native wildlife by some wildlife hating cat owners who either let their cats roam outdoors or simply release them to fend for themselves. Astoundingly, some people think that these unwanted, but now neutered native wildlife killers deserve better than fending for themselves, and create little shrines where equally misguided volunteers bring food for the unwanted cats. It always amazes me how anyone could not realize that 1) neutered cats still kill native wildlife, and 2) people continue to dump off their unwanted cats when they realize that someone else will feed them. TNR does not work, and the only solution for unwanted cats is euthanasia, NOT letting them outside. Remember, however, killing even a feral cat in some states is a felony.

PETA's stance on hunting is another example of the Dunning Kruger effect. For instance, you can read on their website: "Hunting might have been necessary for human survival in prehistoric times...." Unless "might have been" means something else, or you've been living under a rock, how could anyone write this and not embarrass their mothers? How could anyone read this and not consider it an insult to their intelligence?

Or: "This unnecessary, violent form of "entertainment" [read hunting] rips animal families apart and leaves countless animals orphaned or badly injured when hunters miss their targets." Laughable, if it wasn't squarely in the Dunning Kruger zone. We should be unable to walk through the woods and fields without tripping on orphaned, wounded animals. And what animals do not become food for other animals, save perhaps adult elephants or something similar? Should we pity all animal families? Spoiler alert: It's called nature.

PETA is a champion of taking information out of context to fan emotional flames. Consider this: "A member of the Maine BowHunters Alliance estimates that 50 percent of animals who are shot with crossbows are wounded but not killed." Now anyone with any knowledge of this subject knows this is completely, utterly, and ridiculously untrue. Unless it's amended to read "50% of animals who are shot with crossbows from 150 yards at dark, in a thunderstorm with high winds are wounded but not killed" then it might be true. However, it would be 1 of 2, but at least the math would work out.

Back to families: "Hunting also disrupts migration and hibernation patterns and destroys families. For animals such as wolves and geese, who partner for life and live in close knit family units, hunting can devastate entire communities." How long have people hunted? Have geese stopped migrating and bears stopped their winter naps? Since when are families not "torn apart" by natural predation? Oddly, it's ok to let populations get out of control because starvation and disease are "nature's way of ensuring that healthy, strong animals…." Do PETA supporters enjoy watching deer die of starvation or disease rather than via a quick humane kill from a hunter whose family will eat the meat?

Lest we forget to continue the lie about "trophy" hunting: "Hunters, however, strive to kill the animals they would like to hang over the fireplace—usually the largest, most robust animals, who are needed to keep the gene pool strong. This "trophy hunting" often weakens the rest of the species' population: Elephant poaching is believed to have increased the number of tuskless animals in Africa…" Again, how long have people hunted, and have we seen the effects they predict? Of course not, because it's a lie.

It is true, big horn sheep have begun breeding at smaller body size and earlier ages probably due to over-harvest of the largest adult males, but this can be corrected now that it's noticed. And for trophy hunters, PETA ignores the fact that 99% of all deer hanging on some hunters' wall were also eaten, as were the millions of deer that were not mounted by a taxidermist. I can attest to the 1% because a buck I killed with my bow ran maybe 150 yards, but several searchers failed to find it until a week later – the taxidermist said OK, but the meat was barely fit for vultures. Lastly, for elephants, responsible hunters are not poachers.

Although I appreciate PETA's stance against TNR, their take on hunting leads to the unfortunate substitution of emotion for logic and critical thinking in otherwise well intentioned people. We need to do a better job of educating people that nature is an animal eat animal world. Although perhaps antihunters will never understand our motivation for hunting, we need to make clear that hunters are not kill-motivated thrill seekers. Some people, over sensitized by PETA tactics, need rational discussion and not pictures of dead animals. We should post positive outcomes of hunting, from population regulation to the sheer fun of being in a hunting camp or walking fields for pheasants. Or the fabulous venison dinner I just enjoyed.

53. Forensics And The Law

Cabin Talk.—The use of DNA to identify individuals has been a mainstay in the legal system for a couple of decades. The advantage is clear, DNA is not affected by people's perceptions, judgement, and instead is mostly unambiguous. In contrast, even eye witness testimony can be wrong. Below I recount a personal experience from well over a decade ago in which my lab helped catch wildlife violators from a distance.

While at the MSP airport to pick up some incoming specimens, I was chatting with a USFWS wildlife inspector. I asked if they had any interesting birds they had confiscated and she replied that they had some tropical species in their freezer. Of course as a museum curator (Bell Museum, Univ Minnesota), I asked if I could have the birds for our collection, but they were evidence, so no.

The inspector then took me to a freezer in which there were lots of plastic bags with what were obviously completely plucked and cleaned quail and grouse, clearly not waterfowl by color of the meat. The Wildlife Inspector told me that the birds were seized from upland bird hunters returning to the US on a flight from Canada.

USFWS had gotten a tip that the hunters on this flight had over limited on some species and were going to falsify the number of each species on their 3-177 importation form. To do this, they took all the plumage off the birds, against laws governing import. They claimed they didn't know they were supposed to leave a wing attached, and hoped they'd get to keep their birds by apologizing for their (deliberate) oversight.

Wildlife inspectors are not authorized to provide definitive identifications of birds unless plumage remains on the carcass. Hence, the carcasses were seized because they lacked any identifying plumage.

I told the inspector "I can identify the carcasses to species." Surprisingly quickly she told me that I "could not eat evidence in a federal investigation." Too bad, as there were lots of frozen birds all in perfect condition and I thought I could easily identify the species by taste. I then told the inspector that she could bring the carcasses to my lab and we would extract DNA from each one, determine the DNA sequence for a diagnostic genetic marker, and solve the identifications. We were left with a small tissue plug from each bird and she took the rest of the carcasses, again, something about not eating evidence.

It was an easy task to identify each bird because the genetic segment we analyzed provided an unambiguous bar code to species. Our results made it clear that the hunters had falsified their 3-177 import declaration and had clearly over limited on Hungarian partridge, many of which they claimed were other species so as to be within limits on their Canadian hunting permits. But it was obvious to the Wildlife Inspector that there were too many of one kind, even without feathering. I sent the inspector a letter detailing our methods and findings, and sort of forgot about the work.

Quite a while had passed and I wondered what happened, so I asked. I was told that USFWS did not contact the hunters, so the hunters assumed that their plan almost worked. That is, the hunters assumed that by removing all feathering USFWS could not verify identifications of the birds and prosecute them, even if the birds were confiscated (which they were). I was wondering why our information apparently was not used. Now my recollection is hazy here, but the inspector said something to the effect that the hunters believed that inspectors working for USFWS were just plain stupid. That was not the case. Thinking the coast was clear because they hadn't heard from USFWS, the hunters booked a hunt the following year with the same outfitter and flew to Canada. The hunters, however, hadn't counted on forensic identifications using DNA.

Unbeknownst to the hunters, the tip to USFWS at MSP airport the prior year had come from the hunter's Canadian outfitter. Wildlife agents from the two countries shared the genetic information we provided. As the hunters deplaned in Canada, they were met by law enforcement and taken before a judge. Fined for exceeding legal harvest limits the prior year, they were sent back to U.S. I wasn't told whether they lost their guns or hunting privileges in Canada, or how much they were fined. I also do not know if the hunters were charged with falsifying their 3-177 import declaration.

There are a few instances where DNA sequencing could misidentify a bird to species, but in general, the DNA is straightforward, and the identifications clear. It was unambiguous in this case.

Being a hunter myself, I felt a bit guilty for providing the data that resulted in the hunter's convictions, but not too guilty. I have hunted many times in Canada and always followed their regulations. One year when returning to the US, they isolated my 12 yr old son and asked him how many ducks we were bringing back. That's like asking a 12 yr old boy how he does his laundry. So, even though we'd discussed it, he helpfully told USFWS agents couldn't remember how many we had or how many we'd eaten. As a result we were checked to see if my 3-177 declaration matched what was in our coolers. They didn't need forensic DNA analysis because all our birds had wings attached, and we were legal.

In sum, a wildlife inspector is a key player in how USFWS stops poaching and other kinds of wildlife crime Those interested in careers as a wildlife inspector can obtain information at https://www.fws.gov/program/office of law enforcement/get involved. Those wishing to report an illegal activity can do so at https://www.fws.gov/story/how report wildlife crime. To me, this was a fine example of personnel from the U.S. Government and a public university working together. I thank the USFWS for permitting me to recount this story.

54. Check Your Backyard For Dire Wolves

Cabin Talk.—Darkness has fallen, and you're about to get down from your deer stand on a freezing November night. Suddenly out of nowhere you hear an eerie howl very close, in the direction of your truck, and your gut tells you it's not a normal wolf. You'd heard about these reintroduced "extinct" dire wolves that can weight up to 200 lbs with oversized teeth and jaws, that used to eat horses, bison and potentially mammoths. You quickly realize that this howl has to be one of those wolves. Suddenly, given how far away the truck is, your only thought is that you hope that dire wolves can't climb trees. And, you wish you'd have brought more than just your bow. And of course, your cell phone battery is dead.

It is highly unlikely that this could ever happen. At least I hope so. A company called Colossal Biosciences partnered with a high profile scientist Dr. Beth Shapiro who's into "ancient DNA" to reconstitute what they are calling a dire wolf, a species that disappeared from our landscape about 12,000 to 13,000 years ago. At this time, much of the "megafauna" of North America disappeared, including saber-toothed cats, mammoths, giant ground sloths, camels, short-faced bears, and native horses, to name some. Debate continues as to whether this die-off was a result of hunting and food-chain disruption by humans (the so called Pleistocene overkill hypothesis), or climate change that accompanied the end of the last ice age. I tend to lean towards the cause of the extinctions being a bit of both.

The process of creating a dire wolf was not trivial. First, DNA was extracted from a 13,000 year old tooth and a 72,000 year old skull. This sounds easy, but it's tricky because there's very little DNA and it's "degraded" so all the DNA that is harvested are just tiny bits. But these bits can be amplified and superimposed on each other to make a complete copy of the genome. If you recall, DNA is the genetic code that comes in just 4 letters, G, C, A, and T, and it's the sequence of these letters that contains the blueprint for making an animal; DNA is the language of heredity. A super computer and AI look for areas of nonrandom overlap in these short sequences and figures out how to line them up.

When aligned to each other, the gray wolf and dire wolf genomes were 99.5% similar. That sounds like an awful lot of similarity, like maybe gray wolf is just a dire wolf. But if the genomes contain 2 billion bases, the 0.5% difference means that there are millions of bases that differ. However, DNA is organized into genes, which make things, and there are only about 20,000 to 25,000 functional genes in this myriad of DNA bases. The researchers settled on 15 genes that they thought were important differences between dire wolf and gray wolf, and then genetically engineered a gray wolf genome with these changes, put into a dog ovum in which the DNA was removed, and implanted it presumably in a gray wolf to develop.

And voila, three pups have been born, 2 males and 1 female. There are pictures of them all over social media as well as recordings of the young wolves howling. Whether this is what dire wolves sounded like is totally unknown – part of their allure is that they're still puppies, who doesn't love puppies no matter what they do or sound like? Heck, even pug puppies are cute. The newly engineered wolves are in an undisclosed sanctuary where they're looked after night and day. They were hand reared, but some think they'll learn to hunt. However, given that they'll have no parents or pack members to learn from, and no natural prey, will whatever they do be indicative of what dire wolves were really like?

The many social media discussions of these dire-like wolves mostly focus on issues or problems, not the success of the breeding experiment. The first is what I just mentioned, and that is they are not 100% dire wolves, and hence, saying the dire wolf has been "de-extincted" is erroneous. What we have is a GMO that is a gray wolf with some, but not many, of the genetic tweaks that were in the genetic blueprint of dire wolf. One of my paleontologist colleagues remarked "making a GMO wolf larger and white doesn't make it de-extinct."

Second, the habitat in which they roamed, and the species they preyed upon, are long gone. So, there's no way to envision their re-introduction into the wild, an idea called "re-wilding." We have enough issues with gray wolves, and a different species that is 25% larger would create only more problems, although they might kill off gray wolves, who replaced dire wolves after they disappeared.

And consider these newly reconstituted wolves. They'd be an issue, not just because they'd eat deer, but they don't belong in the current ecosystem. It wasn't just the dire wolves and their prey that went extinct, but plant and insect communities have also changed. Our environment today is a totally different place than it was when dire wolves roamed. Are we going to bring back the thousands of now-extinct plants and animals and create a veritable Pleistocene Park?

Because it hasn't belonged to any ecosystem for over 10,000 years, dire wolves might be considered an invasive species. Maybe they'd be listed under the Endangered Species Act, or the Endangered de-Extincted Species Act? I can't begin to imagine what a poached dire-like wolf would bring on the black market. Speaking of invasive species, a colleague mentioned that the dire wolf is part of a South American radiation that "invaded" North America, hence, they're immigrants. Possibly they would need to be deported.

It matters then what you think de-extinction is, and according to Shapiro, how you define a species. If genetically engineering some genes from an extinct wolf into modern wolves creates something that looks different, which is her definition of a species, then de-extinction has been achieved. If that's the case, then a pug and a Dalmatian are unquestionably different species. Most professional biologists, myself included, do not think they have brought dire wolves back from extinction by any definition of species.

Authentic de-extinction would entail making an animal with an entire dire wolf genome, which has not been achieved, and is a ways off. These new pups are a partial replica of a wolf-like animal that lived in a bygone era, and although we might long to know what a dire wolf looked like, we still don't know. Given they can't be released back into the wild, I see this accomplishment as more of a curiosity. However, the technology could be used to recover species that are on the brink of extinction like the red wolf, or species that became extinct more recently, like passenger pigeons. Of course, like the dire wolf, the environment that supported a few billion passenger pigeons is also gone.

The feat that Colossal and Dr. Shapiro accomplished is nothing short of amazing, but no, the dire wolf has not been de-extincted, it is still very much extinct. As a caution, it would be an unparalleled misuse of the science to suggest that we do not need to continue the Endangered Species Act because we can "de-extinct" species, because we cannot. Fortunately, the deer-stand event I envisioned will only be in our (bad) dreams.

55. Are Fish Affected By Boat Noise?

Cabin Talk.—Sitting beside an open window with a cold drink, I see boats racing back and forth across the lake. It just seems to go without thinking that when I'm fishing, I try to avoid areas with lots of boat traffic. I guess I figure that fish think like me, and why not go somewhere to avoid the traffic noise. Actually, pondering this, why I would think that fish think like people? Even if fish are dumb, I would "think" that they'd avoid areas with the most boat traffic.

Noise. We humans contribute noise to that landscape that we almost don't notice. Sitting lakeside, I hear a jet overhead, several boats, a chainsaw in the distance, a truck on a back road, and the neighbor's dogs. And my hearing isn't what it used to be! We try to be quiet when we hunt, so as not to scare away deer or grouse. Shouldn't it be the same for fish?

The noises we produce come in different frequencies, both high and low. We know that animals react to it. For example, some birds change the pitch of their songs along roadsides to avoid overlap in frequency with road noise. Otherwise the information in their songs can be cancelled out by sound waves on the same wavelength as those coming from vehicular traffic. What about things that live in water? Are they subject to human noises?

My grandfather had a cabin on a lake in western Minnesota where I developed a love for fishing. He used to enforce a quiet boat. No loud talking, and if you even hinted at turning on a radio, the world would be ending. Drop something on the floor of the pontoon, there were raised eyebrows and a scowl - you were bad. Same in the fish house in the winter – speak in a soft voice and don't let the door slam. Or what? His opinion was that you'd scare away the fish.

Even today, when I'm racing from one part of the lake to another, I make a wide track around boats that are jigging or trolling, and if I can't, I throttle way back. Seems like the courteous thing to do, and I hope others do it when they pass us while we're fishing. I assume that noise will disrupt their fishing by spooking fish.

Many times I've looked at an area that seemed like it might hold fish only to pass it by because there were lots of boats racing through it. When the jet skis and water ski-ers come out, it's definitely time for lunch. Why? Because they scared all the fish away, grandpa said so. He didn't have an answer, though, for where did they go? Actually, I never thought about that, except that they either went to the bottom and clamped their mouths shut, or they went to some totally different part of the lake or river. Or they too just went inside and watched TV, or whatever the fish equivalent might be.

Once again, all of these opinions are based on what we perceive to be what a fish would do. Now, I've never caught a 20 lb northern when a boat was racing by, but then again, I've never caught a 20 lb northern at any other time. I do think, and probably others agree, that in spring when you're fishing walleyes in clear shallow water, they are spooked by the boat itself, or noises coming from it. With the new technology, forward-facing sonar, my friends who use it say that walleyes are clearly driven away by a boat passing over them (and next to his boat), if the water was less than about 15 feet deep.

Like many things, much of what I "know" I gained by watching and listening to others, who in retrospect knew about as much as I did. How exactly would you know that you weren't catching any fish due to 1) there are no fish there, 2) they're there with lock jaw, 3) there's a bunch of boat traffic, or 4) some combination. Granted, modern electronics can at least tell me if there are fish present, but not how to make them hungry for what's on the end of my line.

I should stop being surprised by what scientists will do, but a paper appeared in 2018 in a top journal (Biology Letters) that tested the effects of motorboat noise on fish behavior, by Harry Harding and colleagues from several institutions in the U.K. Now full disclosure, the tests were done on cichlids in Lake Malawi, Africa, but the details and results were interesting, and hopefully general to other bodies of water.

Different parts of Lake Malawi have different levels of motorboat activity. The researchers identified four low disturbance and four high disturbance sites in the lake, each with same depths, distance to shore, bottom type and water temperature. So the idea is that fish in the first areas rarely experience motorboats, and those in the second set of four sites regularly experience boats going over them. How do they react? Do any fish stay in the high noise areas? I know what my grandpa would say.

You might be wondering how they measured whether the fish responded to noise? It would be pretty subjective to make a video and come up with a value judgement of whether a fish appeared not stressed, kinda stressed, or really freaked out. So instead, the researchers measured oxygen consumption rates! The fish were tested in special underwater opaque containers in which the researchers could measure oxygen consumptions and exclude visual stimuli of boats passing overhead, leaving just motorboat noise. The point? When fish are stressed, they consume more oxygen.

The experiments used three different noises: motorboats passing overhead, speakers playing motorboat noises, and no noise, the last being the all-important scientific control. The control helps insure that the effects noted on the fish are just not a function of being in a container.

Unlike some scientific studies, the results here were fairly easy to predict. In the low disturbance sites, fish consumed more oxygen when they heard motorboats or recordings of motorboats compared to no motorboats. That is, motorboats caused a stressful change in their behavior. At the high disturbance sites, there was no effect on oxygen rates of motorboats relative to ambient conditions! So, given the ecological equivalence of the two types of sites, one could infer that fish become acclimated or accustomed to motorboat noises and don't get stressed out.

My assumption that there would be fewer fish in the high disturbance areas was not addressed by the researchers – that would be a key piece of information. Still, it is interesting to know that fish become accustomed to boat noises, so maybe we should fish those busy channels. This opens up a lot of research questions, one of which is under what conditions freshwater fish in the Northwoods might become accustomed to boat noise: depth and clarity of water, type of noise, frequency of boats passing, etc. In retrospect my grandpa was at least partly right – cut down on noise when fishing your secret, rarely fished, spots!

56. I Got The Lead Out – Please Join Me!

Cabin Talk.—Many of the things we do that harm our environment are hard to change. We've destroyed over 95% of the native prairie for agriculture, but I am fond of eating. I drink diet sodas from a can, although I recycle them. It's never clear what I should do with the plastic bags my groceries come in. I still drive a vehicle that uses gas, at least it's unleaded. But some things we do can be fixed to a great degree. Speaking of lead, I recently made a personal decision about the fishing tackle I use.

I stopped using lead fishing tackle but I hadn't rounded up the lead jigs and split shots in my tackle boxes. What you see in the picture is about 1.5 lbs of lead (hooks cut off). Granted, most of that lead tackle wouldn't have ended up on the bottom of a lake, but I have seen too many images of loons and eagles that have died a horrible death from lead poisoning and decided to draw the line (pun intended) in my tackle box. See https://raptor.umn.edu/about raptors/conservation and research/environmental contaminants/lead poisoning for the latest on eagles poisoned by lead in Minnesota. A vet tech from U Minnesota Raptor Center said: "The pathetic calls of lead poisoned birds are heartbreaking."

Anglers can help fix this problem. We could have a contest to see who can send the most lead fishing tackle to the hazardous waste facility. But lead has been with us for a long time, is it really that bad? The toxic effects of lead on humans, and especially young children, is well known, and the reason we dropped lead from house paints and gasoline. And then there's the ubiquitous lead fishing tackle.

I realize that the amount of lead I discarded is not going to make a difference. But when you recognize that it would take just a fraction of just one of those jigs to kill a loon, it's what motivated me to dump my lead tackle. What if thousands of fishermen and women did the same?

I am willing to wager that ALL of us use about 10% of the tackle we carry around. Get the lead out.

Yes, alternatives are more expensive, but not a deal breaker. I replaced the jigs and weights shown in the picture with tungsten equivalents for about $140, a small amount compared to my rods, reels, lures, line, nets, boat, gas, bait, etc. I got a nice diversity of sizes and colors. Will my gesture save loons as a species? Obviously not, they have plenty of other problems. Would I pay $140 to prevent a loon from an agonizing death from lead poisoning? In a heartbeat.

Some perspective on lead and fishing is warranted. It is estimated that 4,384 tons of lead fishing tackle are lost each year in U.S. waterways. As one example, between 1983 and 2004, roughly 1 million pieces of lead tackle (about 9 tons) were deposited in Mille Lacs, a large walleye lake in central Minnesota that is a walleye destination. Since using lead tackle is not illegal, it's up to each angler to make his or her own decision. For tackle manufacturers, there's a fortune and a half to be made from switching to nontoxic tackle and replacing the lead tackle for those who choose to do so. And the cost to anglers will come down.

There are other lead issues in the sporting world, like bullets. If you did the math, there are probably relatively few lead bullets on a per cubic foot basis in the environment that resulted from big game hunting. But lead fragments remaining in the gut piles are consumed by scavengers like eagles, which should be reason enough to switch to nontoxic bullets. Now, I'm not against killing animals, I've killed my share of deer, squirrels, rabbits, grouse, woodcock, ducks, geese, not to mention fish. But I consider these kills to be humane. There is nothing humane about an eagle dying from lead poisoning. And not to be overlooked, there are many sublethal effects of lead on wildlife species, such as immunosuppression leading to lowered disease resistance.

On a different scale, consider mourning dove hunting. About 14,000,000 are harvested in the U.S. every year, making mourning doves the most harvested game bird in the U.S. Over the last 40 years, that's about 560,000,000 doves harvested. I'm guessing 7.5 shots/bird (probably my average), or 4,200,000,000 lead shells expended onto the landscape, which amounts to 262,500,000 lbs of lead (assuming one ounce of lead per shot). This is over a huge area and 40 years, although some areas have much more lead than others because doves are concentrated. But that's a lot of lead we've put into the environment, and lakes and ponds likely have a long standing base of lead from early days of waterfowl hunting, percolating below the surface

Consider ruffed grouse hunting. Between the years of 1940 and 2009, Minnesota hunters harvested about 34,500,000 birds. Based on my experience, that's an average of about 5 shots for every bird harvested, or 172,500,000 shells, resulting in 10,781,250 lbs of lead that made it into grouse habitat. Ok, I need to subtract the average of 1 pellet I get in a bird, and later bit into, but I have a good dog. It might be trivial on a per cubic foot basis, but still, it seems fair to suggest that the environment would be better off without about 11 million pounds of lead being deposited into the soil and water table.

One might say I picked long periods of grouse and dove hunting to exaggerate the case. But the point is that lead doesn't degrade very rapidly. Although some claim that lead remains unchanged in the soil for 2000 years; that's not the scientific consensus. For example, McLaren and colleagues showed that "Lead was readily released from Pb shot into the soil environment due to rapid corrosion of the Pb shot."

They also remarked "Soils subject to inputs of Pb shot can become highly contaminated with Pb." So, much of the millions of pounds of lead is still there in some form, gradually leaching into the environment, and increasing annually. It is unjustified scientifically to believe otherwise.

The issue with lead shot and waterfowl was that ducks and geese ingest lead, and loons and some water-

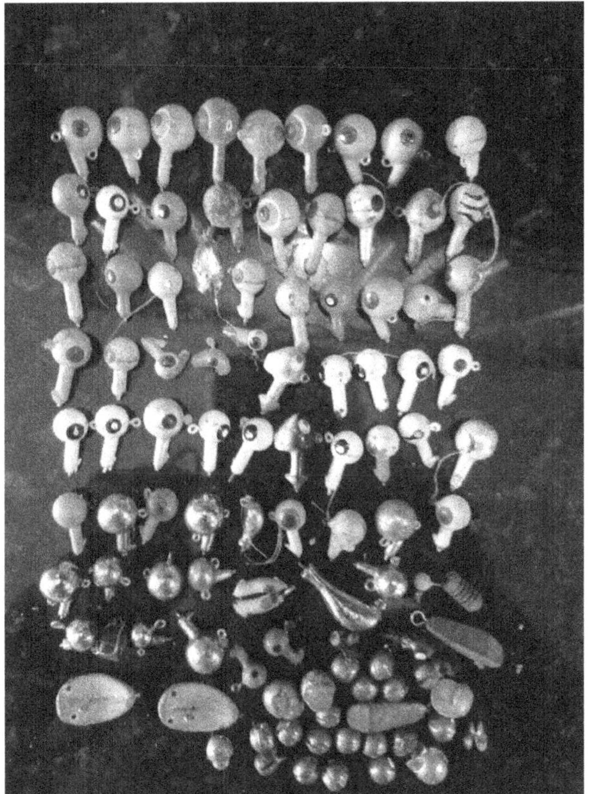

Lead jigs and weights from the author's tackle box headed for the hazardous waste disposal facility. Join me and save a loon.

fowl (e.g., swans) also mistakenly ingest lead tackle. But will several tons of #8 lead shot spread around on the ground, in the grouse forests of Minnesota be a problem? A 2009 study by Bingham and colleagues found that chukars in the western U.S. could not tell a piece of lead shot from a piece of grit, and about 9% of birds had ingested lead pellets. How likely is it that a ruffed grouse will mistakenly eat a piece of shot? A 2005 study in Canada by Rodrigue and colleagues found that 2 of 155 ruffed grouse gizzards had an ingested lead shot. In other words, it might be low so far, but we don't really know.

But the bigger question is, even if few grouse ingest pellets, do lead pellets in the soil of our grouse woods from generations of missed shots at the wily ruffed grouse present a serious contamination issue. Given the scientific consensus that lead leaches into water and soil, I'd guess that all that lead cannot be good. Given the slow degradation of lead in our soil and water and it's continued accumulation, we should consider the future effects it might have, we should switch to something else. We banned lead from waterfowl hunting; we should complete the job.

57. Public Perception Of Hunting, Fishing, Shooting Sports, And Trapping

Cabin Talk.—Hunting has been hit pretty hard in social media. Emotion-laden misconceptions cause some to call for an end to hunting. Hunters have not responded with quite the vigor because it has a long tradition, is legal, and humans evolved as hunters. I personally am a hunter and take pride in providing interesting fare for the table. I wish that those who find hunting objectionable realize that many of us feel differently. If you don't like hunting, then don't be involved with it. If I had to guess from what I see in the media, I would think that hunting is in trouble.

What do the numbers say? A survey entitled "Americans' attitudes towards hunting, fishing, sport shooting and trapping" provides an interesting perspective.

Surveys are tricky, and it's all in how you ask the questions. Many surveys have ambiguous questions that could have multiple different answers. Important too is the wording. "Let's go shoot grandma" without the comma before grandma, has a very different meaning than what (hopefully) was intended. "Have you stopped beating your dog" should elicit no response. A group called Responsive Management conducted pre-tests of the questionnaire to ensure proper wording, flow, and logic in the survey. People study how to make surveys more accurate.

The survey was done via telephone, and there are ton of details in the 89 page report. They tried to standardize the questions and surveyed relatively equal numbers of people in four regions, west, Midwest, southeast, and northeast. Interestingly, given the fact that most hunters and anglers eat what they harvest, the survey did not inquire as to whether the respondents were vegetarians (which represents only 5% of US population).

The survey was under the auspices of the Association of Fish and Wildlife Agencies (AFWA): this is not a government agency despite similarity to USFWS. However, AFWA members include every state agency, BLM, NOAA, NPS, USDA, USFS, USDA, EPA, USFWS, and USGS. Non-government members include AFS, B&C, Delta, DU, NWTF, P&Y, PF, RGS, and Wildlife Soc. Thus, I am not concerned about the survey being tainted by conflicts of interest because of funding sources.

Hunting

Overall, 4 out of 5 Americans approve of legal hunting, with the Midwest having the highest level (86%) and the northeast the least (72%). So anti hunters are outnumbered 4 to 1. This is sobering considering the loud and persistent voices of the 20% who disapprove of hunting.

Some species received higher approval ratings than others, ranging from 78% (deer, wild turkey) to mountain lion (38% approval). The majority (56%) dislike mourning dove hunting, which is ironic because more mourning doves are shot by hunters in the US than any other species. The percent approval varies from region to region for all species, with the biggest regional differences being mourning dove, which range from 32% (Northeast) to 52% (southeast), closely paralleling where most mourning doves live. Over 74% of all Americans disapprove of hunting African Lions. Most attitudes have remained constant between 2006 and 2019.

Stop calling it trophy hunting! Trophy hunting received 29% approval. However, the public is mostly oblivious to what this means. No responsible hunter shoots a big deer and just takes the antlers. The survey inserted "legal" in front of every activity it asked about to eliminate poaching and illegal activities -- just taking the antlers is wanton waste and is poaching not hunting. "Trophies" are also eaten, period.

The survey addressed how game is taken. Bow hunting received an 80% approval, with use of dogs coming in at 55% (doesn't distinguish what sort of game, waterfowl, upland game, or tracking deer), 32% approve hunting over bait, and 21% approve high fence hunts. Bowhunting had strong support in every region and over time, but no attempt was made to distinguish crossbow or vertical bows.

Most of the 80% of the people who approve of hunting don't get their boots dirty and bring home game. About 40% of the US population participated in either hunting, fishing or wildlife related activities, with the majority in the latter category. It is estimated that 10 million people hunted deer in 2017, which amounts to 3% of the US population. So for perspective, although 78% of people approve of deer hunting, only 4% of those who approved actually hunted.

Fishing

Approval of fishing received uniformly high support, 93% overall, with support ranging from 98.5% among hunters to a low of 87.7% among Hispanic or Latino respondents. The question I have is who are the 1.5% of hunters who disapproved of fishing? Maybe they were having bad days. Perhaps we should have dads take kids fishing, because only 91.9% of females approved of fishing. The high numbers were consistent across the four geographic regions. As for methods, many folks approve of catch and release, but only 57% approve of "gigging", which if I had been asked, I would have pleaded ignorance. It's spearing, which in the upper Midwest is a fairly popular way to harvest northern pike in early winter.

Recreational Shooting

A large number of people (81%) approve of legal recreational shooting, with the high end being shooters (98.6%) hunters (98.3%) and anglers (91.2%). The overall highest approval rating was from white males in rural areas who grew up with firearms. I can imagine that the 12% of people who disapprove have never fired a gun, been to a trap range, etc. I've taken many people to a sporting clays range that had never shot a gun and expressed very strong

anti-gun sentiments, who at the end of the trip were dis-
appointed we were out of ammo. The low end (62% approval) was from female, black, urban NE or W and not
growing up with firearms. The word "handgun" did not
appear in the survey, which seemed fairly odd, at least in
this category. I can't go into this, but if 81% approve of
people having guns with which to hunt, why do so many
people, or so it seems, want guns removed from our society? Again, those in the minority speak loudly, because U.S.
gun owners outnumber hunters by 5 to 1, and there are
more guns than people in the U.S.

Trapping

By far, trapping received the lowest approval ratings (52%). Even the infrequent mention of dogs killed in
traps set by irresponsible trappers has a major ripple effect
through the populace. I do nothing in the woods that
could result in the death of your dog. Given how I feel
about my dogs, if they were killed in a trap I'd be furious.
Most instructions on how to remove a dog from a trap are
worthless if your dog has been trapped for even a few
minutes, and you try and wade through the fine print under high stress as your dog is dying. I should not have to
teach myself how to get my dog out of a trap.

Approval of trapping differs among the groups,
ranging from hunters (83.4% approve), shooter (74% approve) and angler (66.2%) to residents of the Northeast
(46.1%), women (42.1%) and African Americans (33.2%).
In terms of reasons for trapping, only when trapping is
part of a restoration program, for subsistence, or wildlife
control did approval near 50%. Unsurprisingly, only 13%
support trapping for fur clothing. Out of the four categories of hunting, fishing and recreational shooting and trapping, trapping is the most precarious according to this sur-

vey. In fact, California recently put a ban on trapping for fur. California officials believe that trapping can lead to overharvest because trappers concentrate their efforts in just a few regions, which might be true there.

Final thoughts

In spite of the relatively high approval ratings of hunting and fishing, participation in hunting by younger people has been dwindling, with fishing remaining relatively stable. The drop offs in numbers of hunters and anglers have real consequences. Fewer participants mean fewer license sales, which results in lower funding for our state game agencies and their ability to manage the resources we use. Fewer hunters and anglers mean reduced sales of gear and equipment, which means less money returned to states from the tax on these items derived from the Pittman Robertson Act. For example, in 2018, Wisconsin received $34,966,603 in Pittman Robertson funds, which support research and other activities of great value to hunters and anglers, and in fact all who enjoy being outdoors.

What should be obvious is that if hunting, angling, and sport shooting are to persist without the kinds of stifling regulations that exist in Europe, we need to get young people outside with fishing poles and guns! The relatively high approval of hunting, fishing and recreational shooting seems to be a green light to rekindle or regrow these activities. High school trap is one obvious success point. Nonetheless, we need to address major issues like access to hunting land, to keep people in the game. Given that the livelihood of game agencies depend on public participation, they need to help renew participation.

58. Forever Chemicals And The Scientific Process: A Tale From Minnesota

Cabin Talk.—I would wager that if you searched news headlines for positive things about how humans have affected our environment, you'd be hard pressed. Perhaps news outlets sell more copy when they report the bad things we've done, but I suspect that the bad outweighs the good. Sometimes it's pretty obvious that the negative impacts on the environment were originally well intended, and once discovered, steps were taken to stop certain activities. One of the major areas in which we've done bad things to the environment involve our lakes and rivers. I have two purposes in the following, explore dangerous forever chemicals in our environment, and sometimes scientific papers go too far.

The world of science can move quickly, especially with new online publication outlets. Still, it takes time to think of a significant question you'd like to answer, design the study, get funding, do the work, analyze the data, write the paper and see it through to publication. The entire process can take years.

Getting a paper published means that it has gone through the gauntlet of peer review, where others in the field examine it to see if there are flaws that undermine the conclusions. There are glitches in the process, but it's the best we have. Many findings in peer reviewed papers are later reversed with new data or better analyses, but that's the way science progresses. If you're a scientist, it's inevitable that something you publish will end up being revised, or worse, refuted. But that's how it works. The important point is that any scientific paper is a stage in the process of understanding.

Contrast that with what we can find on social media, where objective and transparent evaluation of evidence is usually non-existent, mostly hearsay. Sometimes the two ways of communicating knowledge interact in negative ways. Here, I consider a published paper on concerns about eating fish, where science and a sky-is-falling media frenzy was ignited.

I fish because I like eating fish. In fact, I just finished a fabulous breakfast that consisted of gently warmed over northern pike (from Lake Winnie) that was sauteed in panko a couple of nights ago and this morning was atop a bagel topped with capers, cream cheese, and diced red onion. So what's the problem with eating fish?

A concern about eating fish was exposed in a 2023 peer reviewed scientific article in the journal Environmental Research by Nadia Barbo from Duke University and colleagues from the Environmental Working Group (EWG). They presented evidence on the existence of PFAS, or Per- and polyfluoroalkyl substances, in waters and fish. PFAS, or perfluorinated compounds for short, are manufactured chemicals used in products such as food packaging and stain resistant fabrics. They have chemical properties that make them "extremely persistent" in the environment, and are part of over 1000 chemicals previously or currently approved for use in the U.S. One of the main culprits for humans is PFOS, perfluorooctane sulfonate, which contributed 74% of the total PFAS. These are often called "forever chemicals" because of their long term persistence in the environment, which if bad cannot be good.

Fish, meat, fruit and eggs, i.e., diet, is the main way humans are exposed to PFAS. But not to worry, they can also be inhaled via contact with the products and their "dust." Unfortunately, the paper reports that there are tens of thousands of manufacturing plants, landfills and waste water plants, where PFAS get discharged into the water and soil.

What are the health risks of elevated blood serum levels of PFAS or PFOS? Sufficient levels can lead to suppressed immune response, reduced effects of vaccines, possibly elevated cancer risks, increased cholesterol, and reproductive difficulties. These effects will vary as a result of a person's genetics, diet, and overall health. But note, none of the effects are in the helpful category.

Barbo and colleagues reported on PFAS levels in freshwater fish in 501 samples from most states (including MN); data came from National Rivers and Streams Assessment (349) and the Great Lakes Human Health Fish Fillet Tissue Study (152). They found detectable levels of PFAS in all 48 continental U.S. states. Median and total levels of PFOX and PFAS were 2.7 times higher in urban locations. The authors noted that PFAS contamination might be of greatest concern for the Great Lakes ecosystem.

There are serious concerns about their sampling. Apart from 152 samples from the Great Lakes, no other lakes were sampled. This, then, presents a biased view, and we are not sure how contaminated Minnesota lakes might be. For example, the Minnesota fish came from Mississippi River (5), Saint Croix River, Maple River, Minnesota River (2), Big Fork River, North Fork Crow River (2), South Fork Crow River, Des Moines River, Red River of the North, Saint Louis River, Cloquet River, Rainy River, Wild Rice River. The values for PFAS and PFOS for fish from Minnesota rivers were 9.3 and 7.4, relative to national river averages of 20.5 and 16.0, respectively. Major generalizations require strong and even sampling.

Hence, Minnesota river fishes are about 50% as contaminated as the national averages, with higher values from the Mississippi (15.1 and 11.9). A white bass from the Minnesota River showed values of 39 and 33.6, (note to self, don't eat a large white bass from the Minnesota River), whereas a Minnesota River channel catfish had values of 2.6 and 2.6. What is obvious here is that the size/age of the fish was not given and we'd expect older fish to bioaccumulate greater levels.

Here is where the realities of the publication process come into play. The samples analyzed by Barbo and colleagues were collected from 2013 to 2015, but their paper didn't get published for another eight years. The authors mentioned that their results showed a decrease in levels of PFAS from a study just 4 years earlier. This is consistent with the CDC where we learn that from 1999 2000 to 2017 2018, blood PFOS levels declined by about 85%, as PFOS were phased out even though 44% of Americans added more fish to their diets (although much of the new dietary fish is from commercially harvested fish, which show lower contamination levels). But the lag in research to publication results in a need to wonder about whether the results are relevant to current conditions.

These sources of contamination are not new. IBM dropped PFOS in 2010, and 3M announced they were phasing out PFAS in 2022. However, considerable levels remain in the environment. Some places are trying to help. The MPCA is using money from a 2018 settlement with 3M to use new technology to remove these nasty "forever chemicals".

Nonetheless, here is where the EWG group crossed the line from science to fear mongering. They wrote "Eating one bass is equivalent to drinking PFOS tainted water for a month." It should have been "Eating one old large bass from in front of a sewer discharge pipe in the Great Lakes might be equivalent to drinking PFOS tainted water for a month." But that interpretation isn't flashy and won't get social media tongues wagging and has a lesser chance of getting donations to the EWG, one of their main goals. There is no information on the thousands of lakes from which people eat fish, but do they mention this? Nope. The EWG should be on your "watch but fact check thoroughly" list, as they are notorious for fearmongering.

What does this mean for you and me? The CDC considers "high fish consumption" as one fish meal a week or more. For many in the Northwoods, this is met or exceeded at least in the summer months. Fortunately, most of the samples along Lake Superior were at relatively low levels, but again we lack information on Minnesota lakes. The Minnesota Pollution Control Agency (MPCA) tested fish from the St. Croix River, and found that Smallmouth bass, largemouth bass, bluegill, common carp, and freshwater drum all had levels of PFOS that exceeded recommended levels. So even on a local scale, if you're getting fish from contaminated rivers you will accumulate levels of PFOS that exceed the U.S. EPAs interim health advisory levels (for an examplesee https://www.health.state.mn.us/communities/environme nt/fish/docs/menstateguide.pdf). Many rivers are also contaminated with mercury and polychlorinated biphenyls (PCBs).

Given that levels of PFAS are falling, the paper by Barbo and colleagues seems a decade late. But there are lingering levels of PFAS and vigilance would seem prudent, as the MPCA is doing. But so far, according to CDC finding dropping levels, Barbo's maps showing most places with lower levels in 2015, and ongoing cleanup efforts, there is perhaps reason for optimism, at least in most Minnesota waters. Still, valid scientific conclusions await thorough sampling of Minnesota's lakes to determine degree of risk to our health.

59. The Ethics Of Baiting And High Fence Ranch Hunting, A Perennial Debate

Cabin Talk.—What do baiting deer and hunting at high fence ranches have in common? Ethics. Few topics are as guaranteed to start an argument between hunters' with different views of what is ethical. For example, many factions have dug in their heels and deemed baiting to be akin to sleeping with the devil herself. There are some good reasons, both ethical and otherwise. If a white-tailed deer in the latter stages of a Chronic Wasting Disease infection shares a corn pile with uninfected deer, prions (the infectious agent) could be transmitted via saliva to healthy deer, and a CWD epidemic could result, at least in theory. High fenced ranches where hunters shoot what is basically livestock evoke strong negative feelings from those interested only in "fair chase" hunting.

Apart from disease spread, concerns about baiting are largely ethical in nature. Some are concerned that bait piles change deer movement patterns, which they probably do. But so do snowmobile/ATV/hiking/biking trails, roads, subdivisions, new houses, agricultural fields, gardens, nature centers, fences, etc. Many hunters bring up the possibility that anti-hunters will have an even more negative view of hunting if deer are being shot over bait.

It's true that hunters do not need more scrutiny and negative publicity, but I don't think the antis need help.

Not all "bait piles" are the same. A garden, a food plot, a bird feeder, an agricultural field, or a compost heap can legally attract deer away from the natural vegetation upon which they naturally feeding on, to be considered an ethical target. In contrast, many think that food plots are glorified bait piles. This is obviously a slippery slope. We bait bears in most eastern states, but not deer, as it is very difficult to stalk bears and many states use hunting to control bear numbers. On the other hand, some states allow both deer and bear to be baited, some neither. If local game laws reflect what the majority of hunters feel is ethical, it is obvious that hunters (or legislators) in different regions of the country disagree as to what they think constitutes ethical hunting. In principle, sentiments against baiting trace to the well-known North American model of wildlife conservation which provides one definition of fair chase, which explicitly excludes baiting. Given differing view around the country, what is "fair chase" amounts to a subjective judgment as to what is ethical, a perennial problem for human beings.

Another sure way to generate debate topic is hunting at high fenced ranches. In states such as Texas, there are many private fenced ranches where you can hunt native deer as well as a host of exotic species such as feral hog, axis deer, sika deer, red deer, fallow deer, aoudad, blackbuck, goats, and rams of many flavors. You can even save the money you'd need for an African trip and shoot a zebra, wildebeest, oryx or water buffalo in Texas! And to compound our ethical debate, most of these places use bait to lure game into bow or gun range (see photo). Shooting baited animals in a high fenced enclosed ranch sounds like a publicity nightmare for hunters, and surely there must be a consensus that this is not ethical. Obviously, given the large number of successful high fence hunting operations, many consider it "ethical enough" to partake. So do I.

I visited a high fenced ranch in Texas with my two sons starting when they were 10. The first question I get is usually "What is it like, and isn't it like shooting animals in a pen?" Short answer, no. All the animals are "free ranging" but confined in a high fenced area, the size of which depends on the ranch (usually 150 acres and up). Or, conversely, if you're against this activity, you would say that by definition it is unethical hunting because "the animals are trapped in an enclosure from which they cannot escape." This gives the illusion that their harvest has little to do with the hunter's skill, and that you walk around after them until they think you're going to feed them and then shoot them. At a ranch where my son and I bow hunted in Africa, the fenced area was 10,000 acres and there were two female rhinos with calves that wandered the property. So, in this case, I have to think the animals are free roaming, and I suppose if someone wanted to walk around with a bow, I would have to admire their lack of common sense and hope they could outrun an enraged cow rhino.

High fenced hunting ranches usually have a network of dirt roads or trails. After you are dropped off in your hunting spot, ranch personnel drive the roads laying down a thin layer of corn (see photos). Animals typically come to the road and feed on the sparsely distributed kernels of corn. You usually hear the animals first, especially the hogs. In theory they will feed past your shooting window giving you a shot opportunity. So it should be easy to close the deal – just about like shooting them from the corral fence? A perfectly unethical set-up?

This is an oversimplification. The animals are typically wary, especially hogs that have been hunted some-

Top. Road in south Texas on a high fence ranch. Bottom: Close up of same road baited in south Texas on a high fence ranch.

times every other day all year, way in excess of natural hunting pressure. Enough of them know where most of the hunters will be stationed. So if there's even a hint of movement, smell or noise, the whole group vanishes, embarrassed that you were fooled by these "tame" animals. However, the ranches put out feral hogs that were trapped, so they're already wary. It's easy to underestimate these hogs and exotics (except for rams and goats, which are frankly pretty dumb). We wear camouflage, face masks, reduce human scent as much as possible, hide to the best of our abilities, and are as quiet as possible. I'm pretty sure that some deer hunters don't take all the precautions we do at the ranch. Even a skilled bow hunter will find these hunts challenging.

The "hunts" are not a guaranteed deal. Most times I sit without shooting anything. For example, one morning I was in a well-constructed brush blind 18 yards from the road (see photos) and saw nothing. The corn sat untouched – recent rains provided ample new vegetation growth and the animals forgot that they were obliged to visit the road. That afternoon, I was in the same ground blind and about an hour before dark I heard pigs foraging on the road, headed my way. They were active and noisy and from a peak hole in the brush blind I could see them almost leap frogging over each other. I was concealed by a thick tangle of brush and drew back my bow when the lead pigs (usually the smaller ones, or shoats, that the older hogs seem to use as scouts) were about 15 years from my shooting lane. I made no sound, they could not see me, and the wind was in my favor. After a couple of seconds I realized something was wrong – there was now no noise coming from the road. They had veered off about 20 yards from my opening and headed off into the brush – all of them, and then I heard them 20 yards farther down the road, out of range. Hogs 1, hunter 0.

This was really frustrating, as I'm sure they didn't hear, see or smell me. Maybe they detected that someone had recently walked into the blind from the road. However, all was not lost. In about five minutes, another smaller group came from the same direction. Same story – bow pulled back, heart pounding, hogs coming, hogs gone, no explanation. I thought this was supposed to be easy.

My two sons were also hunting. That day, they got one hog between them, a nice boar that my older son (19) shot that evening. A short tracking job was made easy by aid of a fearsome hog-tracking dog named Pickles. Pickles is supposed to find the downed animal and bark. Instead, Pickles usually finds it, takes a sniff and returns to the handler. So, the plan was to "follow the Pickle." Another of the dogs, a diminutive beagle like dog, was named Gator, for his bark not stature.

So much for easy. I had had no shots the first three hunts. I harvested a blackbuck doe the night of the second day and morning of the third. I didn't get another shot opportunity until 10 minutes before dark on the last night (eighth hunt), when just two hogs came down the road. I drew back, and the first hog came up to my shooting lane and then sprinted across the opening, although I am sure he could not see, hear or smell me. He just knew there was a stand there and wasn't taking any chances. Still at full draw, I could see the second one coming but decided to twist around and shoot the first one that was now broadside at 25 yards. I watched my arrow skip under the hog and vanish harmlessly into the brush (we recovered it later and it was a clean miss). "Hog fever" caused me to botch a "gimme." I ended my eight sits with two blackbuck does and a missed hog. My sons each ended up with two animals each. Final tally was six animals for 24 hunts, in other words, we had a one in four chance of harvesting an animal on any given morning or afternoon hunt. This is better than my typical white tail hunts in which it's more like one in a dozen. But it's not a sure thing and nothing like sitting on the corral fence.

Another reason I like these hunts is the opportunity to spend time with my sons, and more importantly, give

View of road from brush blind. My son once found a rattle-snake sharing the same brush pile.

them practice at bow hunting. I consider these hunts as serious "practice hunting." My sons have learned about when to pull back, being quiet, playing the wind, making ethical shots (they have passed many marginal shots), waiting the right length of time to track, etc. They have learned a lot about themselves. I have learned a lot about them. We consider our hunts quality experiences. I do not apologize for taking my sons bow hunting where bait is used. I do not think I have tainted them. They understand the circumstances and that they will usually see fewer "wild" animals.

Some of our other experiences transcend hunting. Once just after dark we picked up my older son, then 15, and learned that he had shot a hog. We were all excited as he'd seen it go down and had already recovered it. As we were riding back to the bunkhouse, he asked me what I thought was a rhetorical question: "Dad, can rattlesnakes hear?" Well, I said they could sense vibration (I've since learned that they can indeed "hear"), but I had a feeling the question was more personal than academic. Sure enough, he told me that after making the shot and seeing where he thought the hog went down, he was going to look for it. As he was about to climb down, he heard

Author's son with a mature blackbuck doe taken with his bow from a ground blind.

something in the leaves under his stand and saw a 6 ft rattler below him about 10 yds away. He yelled at it and it kept coming towards his stand. Then he stomped on stand's ladder and it stopped. Now the story got fuzzy here, but for some reason he lost sight of it, and he thought it had left the immediate vicinity. So he climbed down from his stand and found his hog. My reaction, expletives deleted, was *why* would you get out of your stand with a snake of that size in the immediate vicinity? His response: "You know dad, it's not like it's going to chase me down or something." He wanted to find his hog in the last light and bring it back to the road all by himself. I was "rattled" but concede that he had a point.

On another trip, my youngest (10) was on his first bow hunt to Texas. We stationed his mother in the blind with him, although she's not a hunter, but I felt better leaving him. At the end of an evening's hunt, we arrived at his ground blind and he ran out and exclaimed "Dad! I shot a ram!" He and his mother (his spotter) were pretty sure it went down about 50 yards away. For months prior to the hunt, I had coached him over and over that if he shot an animal, he was to stay in the blind and not risk pushing a wounded animal; he, we and the dogs would all track it. This is standard for bow hunters – unless you see the animal go down, you wait, find and examine the arrow, and then decide whether to track or wait.

I didn't, however, anticipate all possible scenarios. He shot the ram 30 minutes into the 3 hour hunt on a warm afternoon. About an hour after he shot it, vultures started to circle and land in trees above where he and his mom thought the ram had expired. His mother thought they should get out and scare the vultures away. He dutifully reminded his mother of dad's rule that "under no circumstances" can you leave the blind. Vultures landing on his kill was not something I had anticipated. Eventually the vultures flew off because my wife waved her hand out a blind window, over my son's objections. His mother rolled her eyes as my son told me how he had followed my instructions to the letter, but his mother had not. I reconsidered my instructions to "stay put" and amended them to include not only the possibility of vultures, but a rapid exit if a rattlesnake poked its head under the blind. We found the ram right where they thought he had made a great shot.

In summary, my goal has been partly to tell what it is like to hunt on one of the high fence ranches in Texas so that readers can make up their own minds as to whether it's within their realm of ethical. To me, it's exciting and ethical. I am not bothered that the animals come to the road to eat corn, any more than they come to my food plot. These are not wild animals, nor are they tame, and it is hunting to me. I worry about scent and wind, drawing back unseen and unheard, making a clean killing shot, tracking, etc., just like "real" hunting. There is no guarantee you'll get a shot, and when you do, no one tells the animal to stand broadside and quit moving. Any time the outcome is not guaranteed and it is up to you to be in bow range and make an ethical killing shot, it is hunting to me. However, I freely admit that I don't consider these animals to be trophies, but I don't get animals mounted anyway. I figure a few weeks after I'm gone they'll be in a garage sale for $3 each.

Back to ethics. Is shooting animals in an enclosed area over bait ethical? That depends on your point of view, and what you mean by "area." I think we have to realize that what is ethical is in the eye of the hunter — one man's trash is another man's treasure, or something like that. A friend and I visited a ranch in Oklahoma where he arrowed a really nice Texas Dall ram and he had it mounted, although I would not have done so. We can imagine a hypothetical continuum from hunting a grizzly with a pocket knife, surely ethical if not suicidal, to shooting a deer remotely from a computer, which very few would consider ethical. But in between, there is a lot of latitude. For example, what about "hunting" released pheasants on a game farm, does that count as hunting? Does shooting a trophy buck at a food plot remove the "fair chase" label?

I once wrote an article responding to one entitled "Antler Religion" in the Wildlife Society Bulletin. The author railed on about how unethical it was to hunt deer in an enclosed space (size not specified). I think that if an activity is legal, like shooting a deer in an enclosure, then if it offends your personal definition of what is ethical just choose not to partake. When one group of hunters starts telling other groups what is ethical, it's a slippery slope, destined to harm not help hunting.

I am taken by the subtle distinction between different views of what's ethical. On a bow hunting website, a well-known writer, critic of high fenced hunting, praised an outfitter that, on the second morning of making a bow hunting video, got him within 16 yards of an elk. The outfitter wrote on their web page: "We spend numerous hours during the off season locating and tracking elk and mule deer in their natural habitat and continue to track their movement during hunting season to help assure our clients of a true trophy adventure with a high probability of harvest." It was a wild animal harvested legally, but to me it sounds similar to our bow hunts in south Texas, especially the part about the high probability of harvest. It was legal and I think most would consider it ethical, but some for sure would not. I keep coming back to the viewpoint that if it's legal, it's ethical. If it's not to you, then show your opinion by not supporting it.

60. Can Fish Hold Their Breath?

Cabin Talk.—I like to think of myself as a lifelong learner. In an earlier chapter I presented a circle divided up into three parts, 1) things you know, perhaps 2%, 2) things you know you don't know, maybe 2%, and 3) things you don't even know you don't know, or 96%. Life-long learners try to work on 2 and if possible 3, although the task is daunting. I find that many of my :holy cow you got to be kidding me: moments, shortened to holy guacamole, involve things I thought I knew well, and turns out, I'm not even close. Take fish, for example. When my son Matthew, and coauthor of this chapter, sent me a paper about yellow perch, a favorite table fare of ours, I had a holy guacamole moment.

Fish, like us and all other animals, use oxygen, which they get from the water via their gills. It would be pretty interesting if we could dive under the surface and continue breathing. Actually, we have a term for that, drowning.

We have spent many years fishing together, mostly on Winnie and Leech Lake, often targeting yellow perch because they're pretty darn good eating. We recently considered a scientific paper about perch, which showed that they are more interesting than just table fare.

The paper was entitled "Effects of hypolimnetic hypoxia on foraging and distributions of Lake Erie yellow perch," published in 2009 by James J. Roberts and colleagues. It's probably worthwhile to review a few terms from limnology (study of biological, chemical and physical features of lakes and other bodies of fresh water), especially since one of us is a bird guy.

Freshwater lakes in the northern hemisphere undergo annual changes in the distribution of temperature and oxygen throughout the water column, which is referred to as stratification. During summer, a lake has three stacked layers, an upper warmer layer called the epilimnion (about as far down as sunlight can reach), a deeper area of transition called the thermocline (warmer above to cooler below) that prevents mixing with the lower level called the hypolimnion, which is where we find the coolest water. Once the layers are set in the early summer, new oxygen can't enter the hypolimnion, making this comparable to an oxygen tank that won't be refilled until the fall. Meanwhile the top layer can get oxygen from the atmosphere, and stays oxygenated, and therefore more fish friendly.

In more detail, the transfer of heat through water is so slow that the top layer heats up from sunlight and becomes less dense much faster than heat that diffuses

Yellow Perch, image by Pahan

down through the water column, which leads to the layers we observe and is actually why they don't mix, the top layer is warmer and less dense. Almost, but not quite, like oil on water.

In the hypolimnion, dissolved oxygen levels can be very low, sometimes too low for fish. In the fall, with lowered air temperatures, the surface water temperature is lowered to equal the bottom water temps, and with enough wind, the two layers mix, which is termed "fall turnover." During this period, nutrients and oxygen can be spread throughout the water column, and fish follow.

Research has shown that global warming leads to hotter summers and increasingly warmer surface waters, creating a bigger temperature (hence dissolved oxygen) differential with the hypolimnion, resulting in the lake layers stratifying earlier in the spring, mixing later in the fall and stretching our hypolimnetic oxygen tank beyond its capacity. Researchers predict this will negatively affect aquatic lake ecosystems.

Back to yellow perch. Roberts and colleagues investigated the effects of bottom hypoxia (low oxygen) on feeding behavior of yellow perch in Lake Erie. Their working assumption was that the low oxygen zone near the lake bottom would negatively affect perch because although there is potential prey that lives or falls into that zone, perch cannot feed there because there's insufficient dissolved oxygen.

By doing some monitoring, they found that perch do tend to avoid the low oxygen zone near the bottom by either moving away from such areas or moving up in the water column. That is, they can detect and respond to water oxygen concentration. However, and this is the big surprise, they found that perch "dive" into the oxygen depleted zone to feed on benthic invertebrates! How so we asked?

Yellow perch basically "hold their breath" and swim down into the hypoxic zone and eat things like chironomids [midge or lake fly larvae] off the bottom and then swim back out before the lack of oxygen gets them. In other words, perch have the physiological flexibility to get to this food source that is otherwise out of their comfort zone.

We know that many marine mammals also hold their breath, sometimes for long periods – Cuvier's beaked whale can hold its breath for almost 4 hrs! An Emperor penguin for at least 15 minutes. Even insects are able to hold their breaths for considerable periods. But it's hard to think of something stranger than a fish holding its breath underwater. There are, however, a couple of parallels in people. The big breath we take before diving underwater lasts us for some time, similar to how perch use oxygen in their blood to allow shopping trips into the hypoxic zone. Now, exactly what does it mean for perch to hold their breaths? Is it like us when we dive off the dock? Actually, perch just keep pumping water over their gills even though it's not doing them any good in terms of extracting oxygen. So, holding their breath is relative.

Humans have some interesting adaptations as well. Some people in Tibet live at high elevations, over 12,000 feet, where oxygen levels are very low (those who get altitude sickness will appreciate this). The Tibetan adaptation is a simple genetic switch in their hemoglobin gene that allows their hemoglobin molecules to carry sufficient oxygen in their blood even though it's in very low concentration at that elevation. It's almost like holding your breath underwater.

What then will happen to fish like perch with rising global temperatures causing greater oxygen stress in our waters? No one knows. For one thing, 99% of species that ever existed are extinct, which we can probably chalk that up to an inability to adapt to changing climate conditions (new predators, competitors and diseases also have roles). On the other hand, we have seen short term adaptations to changing environments in some species. If perch can withstand low oxygen areas or happen upon some lucky genetic mutation such as that found in the Tibetans, perhaps perch can avoid being in the 99% group.

For now, at least in many lakes, perch seem a long way from blipping out evolutionarily, and it will ultimately depend on how fast the aquatic environment changes and their ability to adapt.

61. The War Over Coyotes

Cabin Talk.—Few creatures are more polarizing that coyotes, love em or hate em with little middle ground. I admit that a coyote in full pelage on a winter's day is a striking creature. I admit that when dog sitting our son's pug, and it decided one evening to charge out (off leash) and challenge a coyote that was calling about 50 yards away, the near fatal encounter was pretty chilling. My voice screaming at a volume I didn't realize was possible seemed to distract both coyote and pug long enough to grab the latter and thereby avoid having to tell my son that a coyote killed their pug.

Coyotes figure prominently in the social media groups I belong to. The opinions are, to say the least, polarized. People who have lost cats, small dogs or poultry to coyotes are usually anti-yote. Frankly, any coyote that kills a cat outdoors is my hero. On a group devoted to crossbow hunting (mostly deer), the opinion of 95% of the writers is "kill every one of them" and the contributors state as fact that increased numbers of coyotes (or wolves in other areas) explain why they see too few deer. Mike Scott's Buckmaster's web page reports that a trail camera outside a coyote den in Vermont recorded 27 fawns brought in over a four week period. Other accounts tell of 17 or 20 fawns brought to a single den. Disclaimer: these numbers were not verified. If true, coyotes could wipe out

Coyote, image Veronika Andrews

deer recruitment in your neighborhood. Remember, though, the chances of a fawn reaching maturity are pretty slim even without coyotes. No matter, most deer hunters want coyotes gone.

On other social media, the verbiage of those opposing coyote hunts is downright hostile: "Let's call coyote hunts what they are: mindless bloodbaths." California, New Mexico, Vermont and other states have outlawed predator killing contests. In Minnesota, coyotes are not regulated and populations appear unaffected despite a number of coyote hunt contests. Coyote hunts violate my personal ethics because I try to stick to the idea that you eat what you shoot, and coyote isn't on my menu. But where legal, it's your right.

Ted Williams wrote a nice summary in 2018 of why predator killing contests harm the perception of hunters in "YaleEnvironment360". If you're killing a pile of coyotes and leaving them to rot, and it doesn't affect their populations or deer numbers, you're not doing hunting a favor.

Others dislike persecution of coyotes for different reasons. A farmer commented "I don't shoot coyotes. I have cattle but have never had a problem with them and they eat fawns so a good deal. I have far more problems with deer." Another commented: "The coyotes are my best friends. I tell everybody shoot all the deer u can on my farm, shoot a coyote and u will never be allowed back."

Coyote populations fluctuate like anything else. Too many coyotes leads to reduced fox populations, which allows rodent populations to increase, resulting in more deer ticks and greater Lyme prevalence. Yes, coyotes eat rodents, but they're not as good at it as foxes.

Waterfowl biologists sometimes control coyotes to increase duck nesting success, although it's likely that foxes do more harm to duck nests that coyotes, so leaving coyotes alone might actually enhance nesting success. Also, coyotes tend to drive off other duck nest predators, such as mink and raccoons.

The coyote was once a western species, and the cutting of eastern forests for agriculture allowed the influx of coyotes (see map) and many other species. Coyotes have largely replaced former large predators, such as wolf, black bear and cougar. Wolves kept coyote numbers in check. Many advocates of killing all coyotes to enhance deer numbers come from the eastern U.S. You cannot admonish them by saying "Coyotes have always been part of the ecosystem" because it's not true. Is there any science to back this up?

Eugenia Bragina and colleagues published a study of deer and coyote populations in eastern North America in the Journal of Wildlife Management in 2019. They compared estimates of deer and coyote abundance in six states and 384 counties from 1981 to 2014. They predicted that deer populations would decrease after newly established coyote populations reached sufficient size to pose a threat. They reached the opposite conclusion: "coyotes are not controlling deer populations at a large spatial scale in eastern North America."

The Bragina study was critiqued in the same journal by John Kilgo and colleagues. They reached a different conclusion when they restricted their analyses to more local scales. They conclude that the Bragina study erred by averaging across studies from different areas. One can see on the map that the percentage of fawn mortality owing to coyotes ranges dramatically across the landscape from 1% (Michigan) to 62% (South Carolina). It is worth remembering that a large percentage of fawns die before reaching maturity, so coyotes (and bears) might be taking many of those that would have died anyway. Hunters kill the largest percentage of adult deer.

If you look across the eastern U.S., many of the northern localities show relatively low percentages, and they tend to be west of the dashed line, which shows the boundary of coyote range around 1900. Coyotes arrived in South Carolina in the 1980s. Perhaps in areas where coyotes and deer have coexisted for longer periods, coyotes are less efficient fawn predators because deer have become better and "outfoxing" them. In the relatively recently colonized southeast, coyotes might well be limiting deer numbers through fawn removal. However, other factors cause deer numbers to fluctuate, including antlerless harvest, where killing a doe might kill two fawns for each year the doe might have lived. Kilgo and colleagues concluded that at the local level, "coyote predation represents an important influence on deer populations." That is, both hunting and coyotes cause local population declines in deer.

Will culling coyote populations help deer? There are two widespread notions about coyotes. Coyotes howl at night, and one of the consequences is that howling allows packs to determine how many coyotes are in the local population. When you break up packs by shooting, the remaining coyotes form more packs than existed before the cull. Second, after a cull, remaining females start breeding at an earlier age; basic population biology is clear,

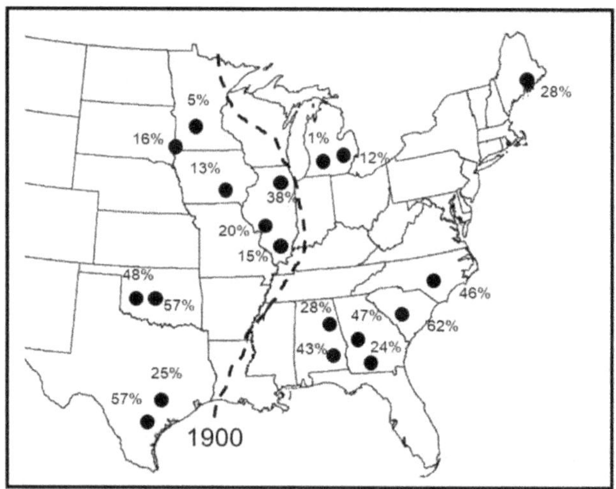

Sites of studies showing percentage of fawns taken by coyotes from Kilgo et al. 2019, J. Wild. Mgmt. Dashed line indicates coyote distribution in 1900.

reducing age at first reproduction makes populations grow much faster. Therefore killing coyotes could lead to more and not fewer coyotes. There is a "carrying capacity" for coyotes (and all things) and earlier breeding by females might rapidly refill the habitat post culling. Peer reviewed scientific papers address these factors.

A 2017 paper by Kierepka and colleagues in the Journal of Wildlife Management concluded that both breeding by younger females and immigration of coyotes from outside populations allowed local coyote populations to recover. In a different paper, Kilgo and colleagues contradicted the idea that culling resulted in increased early breeding: "we observed only weak evidence for compensatory reproduction in response to trapping pressure" and that the re-establishment of the population "was attributable primarily to immigration from neighboring areas." In other words, reproduction by younger female coyotes following a cull might not be the main factor in re-establishment of coyote populations. Furthermore, it is not clear that there are more coyotes than there were before the cull. In fact, Dr. Kilgo, emailed me that "I am not aware of any research demonstrating this." His statement was backed up by a study that showed that there are two kinds of coyotes: residents and transients, with the latter comprising about 35% of all coyotes. Transients quickly fill any vacancy caused by killing a resident.

Large scale culling in the southeast probably increases local fawn survival over the short term, but it has to be done continually and over a large area, neither of which are cost effective. Going back to basics, limiting antlerless harvest could compensate for coyote depredation, because it's the only controllable factor – South Carolina deer hunters killed over 28,000 coyotes in 2016 but didn't reduce the population. Some might argue you need both. The South Carolina DNR states that "overtime [sic], coyote populations are expected to stabilize allowing deer, turkey and small game to still exist in healthy numbers," much like they seem to have done in areas where coyotes and deer have coexisted for a longer period.

The bottom line is this. It's incorrect to automatically assume that if you kill a coyote, you'll save a fawn, although it's possible. Don't shoot a doe and you might save two or more fawns. If your deer population is large, harvesting does isn't a problem. If your deer population is declining, you need to curtail doe harvest, as is SOP when trying to grow herd numbers. In any event, the war over coyotes is far from over.

62. Fish Slots: Sound Science, Or Arbitrary And Capricious Management Decisions?

Cabin Talk.—In many lakes there are limits for the sizes of game fish an angler can keep, and these differ between lakes. Minnesota provides several examples. For example, on Winnibigoshish you can keep (and possess) 6 walleyes, one over 23 inches (tail pinched) or all under 18, whereas on Leech Lake, a few miles away, the limit is 4, with one over 20. Of course, you're on the honor system. If I have 6 walleyes in my freezer at my place on Leech Lake, I'd have to say they came from Winnie. It's possible that walleyes from the two lakes could be told apart because I have preserved tissue samples in my museum (Univ. Nebraska Lincoln) that could be compared genetically.

Where do slots come from and why does it matter? I'm told that if you do a fly in trip to a Canadian lake and catch a few hundred walleyes in a week you'll know what fishing was once like in the Northwoods, long ago.

Fish managers have the tough assignment of trying to preserve as good an experience as they can while keeping populations at a level that will sustain some level of harvest. I'm all in favor of attach and release, if I have enough fish to eat.

Setting fish limits requires information on the number of females in each age class, how many eggs females in each age class produce, and how long fish live. The longer a female lives, the more eggs she will contribute over her lifetime, with egg numbers and egg quality increasing with each year she lives. A 28 inch walleye female can be 15^+ years of age, and she will produce far more eggs than a 20 inch walleye, which might be 5 - 7 years old. Incidentally, spawning females lay 200-400 eggs at a time, at about 5 min intervals, which are fertilized by one or more males.

Surveys provide baseline information for setting limits and slots by documenting the distribution of sizes that reflect ages (see chart). At various intervals, there is a "breakout" year class, or a very successful recruitment year that can be detected with the size distribution (see chart, size 7-8 in fish). The goal is to protect the size classes of females that produce the most eggs, and enough smaller fish to reach the size classes of female walleye with the best reproductive potential. However, these are not the biggest females, simply because there aren't that many of them (chart). So, even though a 20 inch walleye produces fewer eggs than a 28 inch walleye, there are far more 20 inchers and in sum the 20 inch fish make the greatest contribution to the total number of eggs.

Furthermore, if walleyes in the 18-23 inch size range are the most abundant, those are the ones you're most likely to catch. Hence, these are the fish you want to protect with a slot. I once had someone tell me he should be able to keep 4 walleyes no matter their size. Obviously, this would quickly deplete the main contributors to the egg pool.

A brief introduction to the science behind fisheries management comes from NOAA Fisheries. Their first step calls for estimating the "overfishing limit." This is defined as the level of catch that is at the population's maximum sustainable yield. If the catch level exceeds this limit, it will jeopardize the population's capacity to maintain a maximum sustainable yield.

A fisheries management agency sets an "acceptable" biological catch, which is less than the overfishing benchmark. This lowered level is to account for "scientific uncertainty," a common sense management approach. This includes uncertainty about the actual population size/age numbers, and other sources of mortality, prey abundance, reduced habitat owing to greater water clarity because of zebra mussels, spiny water fleas, Eurasian watermilfoil, global warming, or fish diseases, all of which change over time. Even the Clean Water Act is thought to have been hard on walleyes. Hence, it's best to throw in a dose of caution along with the knowns, including hooking mortality (7-16% depending on temperatures).

I've often asked myself, how many walleyes are in a given lake at this very moment; there has to be an exact number, excluding any fish not in the process of being swallowed by another fish. Unfortunately, estimating population size is time consuming and expensive, especially on large lakes. Fisheries biologists use electrofishing to stun fish, mark them, and after they recover, let them go. Next, biologists sample the same area at a later date and compare the number of marked:unmarked fish caught. For example, in one year on Minnesota's Mille Lacs, fisheries personnel marked and released 20,000 walleyes, and later they recaptured 154 tagged and 6,500 untagged. The basic formula is: population size = (number marked x number captured on second visit)/number marked fish captured on second visit.

This led to an estimate of 727,000 walleyes longer than 14 inches in 2018, with 400,000 being females. This doesn't follow from the formula above because corrections were made for capture differences between sexes. In addition, there is an estimation error, meaning it is actually 727,000 +/ some unknown number, which is likely fairly significant. During the period from 2002 to 2018, their six estimates ranged from 1.1 million in 2002 to 249,000 in 2014, showing a 4.5 fold difference in population size. That magnitude of difference is almost certainly valid. Obviously, we'd like to know exactly how many walleyes of every size class are in the lake, but this is a difficult goal to achieve.

These numbers should cause anyone who criticizes

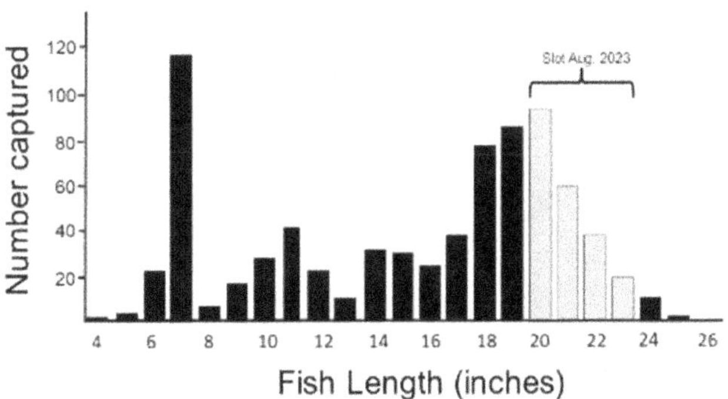

Fall gill net survey Mille Lacs 2022. Minnesota Department of Natural Resources.

fisheries biologists to realize this is a truly difficult task requiring study over many years. Setting limits and slots requires information that is expensive to gather and has

inherent variability. Fisheries managers use established guidelines, with scientific fudge factors, to set annual
catch limits (including take by Native Americans). If trained fisheries scientists cannot always accurately set limits and slots to maintain acceptable harvest levels, it is unlikely that a few friends and I standing around the water cooler after a weekend fishing outing could do better.

What is the goal for a fishery? A good example that would apply to any lake comes from data gathered in Minnesota. Fisheries biologists assume catch rate is about 0.25 fish per hour per person. In Mille Lacs a couple of years ago, the biologically deemed safe harvest of 150,000 lbs was reached early, and by the end of the season, it was estimated that 1.2 million pounds were caught and released, with an estimated catch rate of 0.8. Obviously, if all these fish were harvested, it would have devastated the adult population. If you want a catch rate of 0.25 fish per hr, on average, then build a fisheries management plan to achieve that goal. It won't be easy and expect to miss the goal more often than you'd want. I hope reading this increases appreciation for the dedicated fisheries biologists out there trying to protect our resources.

63. Even Vegetarians Aren't Safe From Bear Meat Stew

Cabin Talk.—I have friends who have varying degrees of meat aversion. One will eat fish, but no other meat. One has a pet pig and will eat nothing with pork in it. Another is a "locavore." Some are vegetarians, and I know couple of vegans. And then there is part of my family, whose reaction to game meat is "ish." No hope for them.

A report circulated widely in the media claimed incorrectly that some hunters contracted a transmissible spongiform encephalopathy like variant Jakob Creutzfeldt Disease from eating deer infected with CWD. The report, although alarming, was not proof that CWD had crossed the "species barrier" from deer to humans. Now, we have a report with far more solid evidence that people were sickened by a parasite after eating bear meat.

In July 2022, a man was hospitalized with a bunch of nasty symptoms: fever, severe myalgias, periorbital edema, eosinophilia, and other laboratory abnormalities. This man had recently been at a family gathering in which bear meat had been eaten by 8 people. The meat from a bear harvested in Saskatchewan had been frozen for 45 days before thawed and cooked. Apparently, guests didn't recognize at first that the meat was really rare and returned it for some more cooking. All seemed well.

Until later, that is. Six of the 8 guests came down with symptoms later confirmed to have been caused by ingesting *Trichinella*, a type of parasitic nematode. Two of the 6 (women aged 29 and 54) had not eaten any meat, only the vegetables that had been cooked with the meat. Being vegetarian didn't help in this case. Because of the relative rarity of this type of infection, two of the hospitalized sought care multiple times before the cause of their illness was identified. FYI, treatment is with the anti-parasite drug albendazole.

The proof in the pudding, or stew, came when remaining frozen samples of the bear meat were sent to the CDC. There are blood cues that indicate *Trichinella* infection, but the coup de gras was that even after 110 days in the home freezer (temperature unknown) microscopic examination of the meat revealed motile (living) *Trichinella* larvae! The exact species was identified as *Trichinella nativa*. Normally, sufficient freezing kills various species of *Trichi-*

nella, but unfortunately for some people, there are freeze resistant species (including *T. nativa*).

These parasitic nematodes occur in many other wild animals, including polar bear, wolverine and cougar. The frequency of infection in black bears from Canada and Alaska ranges from 1% to 24%, so in some areas the chance of coming home with infected meat can be as high as 1 in 4 bears.

Trichinellosis is uncommon in the US, in large part because of the changes in pork production methods. Between January 2016 and December 2022 there were 35 probable and confirmed cases of *Trichinella* infections reported to the CDC and bear meat was suspected in most. Note to self, cook it well! And be careful handling the raw meat. The CDC recommends cooking to at least 165° F to kill *Trichinella*. And because there are some non-specific symptoms, and it's difficult to detect in the early stages, if you feel ill after eating bear meat, get checked out and inform medical staff that you recently ate bear meat.

64. Can Hunters Save The Arctic From Snow Geese?

Cabin Talk.—It's pretty uncommon for hunters to be chastised for not killing enough animals. I've been in a few conservations with non-hunters over my hunting activities, which usually end like this: Me: are you a vegetarian? Them: No, but there is no reason to hunt. Me: every time you buy chicken, beef, pork, fish, whatever, part of the price you paid for that meat goes to the person that killed that animal for you; I'd rather do it myself and know how the meat was handled. As I once remarked to a neighbor when we were discussing a topic where we each had different views, some people see the doughnut, some people see the hole.

Actually I want to write about snow geese and the attempt to control them through the spring "conservation order." About 15 years ago when the spring season first opened, some friends and I went to the area around Yankton, SD. We put out some decoys, and other times crawled through ditches, and made a few rushes, ending up with around 75 snow geese. Not bad considering we'd have none without a season, and we were pleased. Although snow geese are often considered bad table fare, spring birds are quite good. Remember the adage, you are what you eat. These spring birds have been eating spilled grain all winter long.

The next couple of years we did worse and worse, as the birds got smarter and smarter. You could see that flock descent to a field had changed, instead of coming in low, they were staying high and going almost straight down, avoiding roads, and especially any moving or parked vehicles. I know from taking my ornithology classes to witness the snow goose spring spectacle that that was new behavior, quickly learned to avoid hunters.

Our spring hunting ended when it became harder to find hunting spots that didn't already have hunters, and when we finally did find a place, we set up all of our decoys, left after dark and arrived before first light the next morning. Every one of our decoys had been stolen during the night. We went home and didn't return.

Others, however, have continued to hunt the spring conservation order. I'm sure that game agencies invented this terminology to avoid, if possible, conflicts with anti-hunters and those that think the springtime should be sacred because birds are breeding. However, the rationale for wanting to thin the snow goose herd is clear and compelling.

Snow geese are literally destroying much of their arctic tundra breeding grounds. One might ask, why is this happening just now? The answer is that because of lots of waste grain in the wintering areas, snow goose populations have burgeoned, because of the increase in over winter survival resulting from an abundant food supply. In fact, populations increased at an annual rate of 9% since the 1970s! Snow geese in low numbers obviously existed in arctic regions without causing irreparable damage at times (when there weren't large glaciers extending down into the U.S.). But because of the way they forage, pulling out the entire shoots of the plants they eat, when they are overabundant, they prevent the habitat from regenerating. This habitat destruction has affected many other species as well, including other plants, insects and birds.

The ecological devastation caused by overabundant snow goose populations was recognized a couple of decades ago, and the spring season was opened to reduce the populations. It came with liberalized bag limits, use of electronic calls, and use of unplugged shotguns (and even magazine extenders that gave some guns 8 or 9 shot capacity). My 94 yr old cousin said they used to tape a few sheets of toilet paper to a decoy or corn stock and they served as well as decoys. We evolved to using rag decoys, and then many outfitters and individuals with deep pockets started buying up large numbers of full bodied decoys. Given that the season can be short, many of us tried to not break the annual hunting budget on decoys.

I admit that I have recently begun hunting snows in the spring again. The last time in central Nebraska, the group I was with had about 800 decoys out, lots of rags, but quite a few full bodies. It was painfully obvious that the geese knew this gig and as a flock would head for us, losing altitude and then turn at a right angle and fly off, well out of range. We got about 25, most immatures. But it was stark to me to see how easily they recognized a decoy spread and the entire flock turned away.

It's pretty easy to understand why. If a snow goose lives for 20 years, and they are hunted from Sept 1 to mid-spring, then they are under fire for 75% of their lives. They are smart and have learned. A new paper in the Condor by David Koons and colleagues shows just how well the birds have learned, and conversely, just how much the hunters have not kept up.

The authors had shown previously that the mid-continent population of geese continued to grow despite the liberalized spring hunting season. In the new analysis, they used data from 102,600 females marked and released and then resighted or harvested between 1969 and 2016. A sophisticated mathematical model was applied to these mark recapture data. The harvest rates for adult snow geese were highest in the 1970s and have stabilized at the lowest levels observed over the last 48 years, about 3%! Over the last 10 years, adult annual survival probability averaged 84%, which is very high. The numbers don't sum to 100% because there is non-hunting mortality.

We all know that juvie geese aren't as savvy as the older birds. The data are consistent with that, in that survival of hatch year birds averaged only 12% in the recent past, although not all mortality was owing to hunters. In fact, harvest rates of young birds has actually declined since implementation of the spring season. You might think this alone would result in a crash of the overall population. But with adults living a long time, it's not much of an effect after all.

Koons and colleagues concluded, "…harvest has become satiated by the sheer abundance of mid-continent Snow Geese and is unfortunately having a negligible impact on mortality in our study population and those farther north." That is, hunter numbers have seemed to stabilize but the goose population keeps growing.

What's in store? As the mid-continent population expands, it is encountering areas with different plant species that might not be as favorable, although the authors noted that a warming climate might still allow further growth of the population. The authors suggest that game agencies need to "refine and enhance existing harvest conservation measures until efforts begin to increase harvest rates, decrease recruitment rates, and decrease the abundance of light geese." Another option is to allow sales of snow goose meat, which might serve as an extra incentive.

My own view is this. Snow goose hunters have upped the ante with the many gadgets we can use, but the geese have seen them all. I think we've exhausted the decoys and calls. I bet that snow geese fly up to a decoy spread and say among themselves, "Hey, that's the e caller from Cabela's, isn't it?" I think that game agencies should at least explore the possibility of letting hunters use live decoys. Yes, I'd head to a likely spot and put out a few decoys, or maybe a lot, and put up a small obscure pen with some live white tame geese, and maybe I'd paint some black wingtips on them. I'd like to see if snow geese would come down to the sight of real white geese. Now, I understand that I have to maintain a flock of geese at least during the spring season.

65. Should We Hunt Swans?

Cabin Talk.—At the heart of most issues are two opposing viewpoints, and despite the potential, relatively little middle ground. Throughout many areas there are polarized views on whether waterfowl hunters should be allowed to target swans, the sheer image of which tugs at the heartstrings of environmentalists. Why swans and not mallards? Swans are treated to superlatives like majestic and practically, hunting once reduced their numbers to critical levels. And in some areas, trumpeter swan populations have recovered nicely because of intensive and expensive restoration efforts. Obviously, those responsible for these efforts are not enthusiastic about hunting swans! So where are we today?

You and a friend are in a duck blind in western Minnesota and just after dawn, two large white birds fly over your decoys, your buddy whispers "Snow Geese, get ready." You both jump up and each drop one. Perhaps you didn't notice the long necks and that the wings were entirely white, not with black wing tips like snow geese. Most likely, the guys in the neighboring blinds are already on their cell phones to the TIP line at the DNR turning you in for shooting swans. Just how much trouble are you in?

At present, quite a bit. There is no swan season in many states, even for tundra swans, which congregate each fall by the thousands on big lakes and rivers. However, there are seasons on tundra swans in nine US states (Montana, Nevada, Utah, South Dakota, North Dakota, Alaska, Virginia, North Carolina, Delaware). Idaho is considering a season. In Utah, the swan harvest is monitored and after a certain number of trumpeter swans are taken, presumably accidentally, the season closes. Swans, at least trumpeter swans, were nearly hunted to extinction because they are great eating. The only swan (tundra) I've eaten was excellent, right up there with Sandhill Crane.

In August 2019, the USFWS issued a final environmental assessment entitled "Proposal to establish a framework for general swan hunting seasons in the Atlantic, Mississippi, and Central flyways." Unfortunately, not too many swan enthusiasts read it. The document states: "The purpose of this proposed action is to establish a framework for hunting regulations to govern the take of both trumpeter and tundra swans in the portions of the Atlantic, Mississippi, and Central Flyways that currently have operational hunting seasons on EP tundra swans or may have in the future. This will allow limited take of trumpeter swans, but only during hunting seasons established to provide opportunities to hunt tundra swans." Because trumpeter swans are expanding into places where tundra swan hunting is allowed, a primary purpose of this environmental assessment was to limit liability for hunters who might accidentally shoot a trumpeter swan. It was not a call for a trumpeter swan season, despite many erroneous clams To the contrary by conservationists.

A burst of activity among people opposed to swan hunting occurred on social media with the USFWS publication in the Federal Register on March 19, 2020 entitled "Migratory Bird Hunting; Proposed 2020–21 Frameworks for Migratory Bird Hunting Regulations". I initially assumed that it was about hunting tundra swans, but the framework includes potential take of trumpeter swans building off the August 2019 document. The March 19 USFWS posting gave little time for public responses (comments were due April 20), which further enraged the swan loving public (I like swans too). The point of this second document was "to allow State selections of seasons and limits and to allow harvest at levels compatible with migratory game bird population status and habitat conditions." It dealt with ducks, geese, mergansers, sandhill cranes, coots and rails, snipe, woodcock, doves, American woodcock, and swans.

Why would the USFWS open a swan season? The answer to this question is where the root of the conflict between the two sides begins. In a nutshell, unlike many comments on social media, USFWS was not opening a swan season because it is up to individual states and flyways to decide whether to have swan seasons. The USFWS job is to provide a framework so that if a state decided to have a season, there would be guidelines in place. For states with current swan seasons, the USFWS recommended that: "hunting permits for Eastern population of tundra swans be reduced from 12,000 to 9,600, with 5,600 permits allowed in the Atlantic Flyway and 4,000 permits allowed in the Central Flyway." Discerning readers will note that there was no mention of any number of swans proposed to be taken in the entire Mississippi Flyway (none of the 9 states with swan seasons are in the Mississippi Flyway), showing consideration by USFWS for the recovery efforts aimed (no pun intended) at swans.

The misinformation disseminated on social media resulted in a huge public outcry. For a local example, people were told on social media that trumpeter swan hunting would be opened in Minnesota. You can read the 414 responses at https://www.regulations.gov/docket?D=FWS HQ MB 2019 0004. Given that Minnesota has spent a ton of money restoring trumpeter swan populations, the reaction was predictably almost totally negative. Some hunters wrote in and most also disapproved of swan hunting. Few of the letter writers noticed that it was not a trumpeter swan season that was being proposed, but a general swan season, with special attention to limiting take of trumpeters. Many of the letters were downright insulting to the great biologists that work for the USFWS.

According to the MN DNR, there are currently 30,000 trumpeter swans in Minnesota, a wonderful restoration success story. No one wants to degrade the success of the reestablishment of trumpeter swans. However, we can recall the time when the giant Canada goose was restored. Once near extinction, they were subjected to restoration efforts, and within a short time they rather quickly became pests, major nuisances on beaches and golf courses. Other states like Nebraska that got excess geese from Minnesota quickly said "no mas" as they had become pests in their states too. Today MN DNR spends money to reduce goose numbers, have an early season, etc. I guess that's positive problem, so a future swan season is certainly thinkable.

I decided to look at the nitty gritty details of a swan season. USFWS framework says would happen in newly proposed swan hunting areas: "New swan hunting seasons (i.e., seasons in areas that are currently closed to swan hunting) will not be approved unless the requesting State demonstrates that > 90% of the swans in the proposed hunt area are tundra swans." Ok, check. Then, what exactly could the damage be? With 30,000 birds being less than 10% of the total swan numbers, by chance hunters would have at most a 10% probability of accidentally shooting a trumpeter swan. If the number of permits were, say, 2000, that could mean a maximum incidental take of 200 trumpeter swans, less than 1% of the total population. Swans are long lived although each pair has 5 7 young annually (beginning at age 3 or 4), but 200 a year might be-

come an issue if the population decreased for some reason.

The most disturbing aspect of this to me is the misinformation circulated to the public. No mention was made of what the notices in the Federal Register really said, and no mention of how a swan season could affect trumpeters. I see this as more than a lack of communication, it represented both sides choosing not to listen to the other. I think I could tell a trumpeter from a tundra most of the time, but they can be confusing if you see just a couple of birds. Tundra's are usually in big groups, and if they vocalize, it's hard to confuse the two. Additionally, the Trumpeter swan weighs almost twice what a tundra swan does, but if you just see one bird, that's a surprisingly hard call to make. I can see how people get upset about swan hunting, but fact free emotional outbursts do not substitute for civil discussion about wildlife management.

66. If It Looks Like A Mallard, Quacks Like A Mallard, It Might Be A Mallard

Cabin Talk.—Although I have an earned Ph.D. and have taught for 40 years, I am constantly reminded that I missed a lot in my classical education. For example, I have heard the phrase used in my title many times but never understood its derivation. Turns out it's a form of abductive reasoning, not to be confused with deductive reasoning, and it means what you'd expect from the saying about ducks. It's the most likely conclusion one would draw from a set of observations, in this case about ducks. However, in this case, there's a twist.

In the years 1999 to 2001, waterfowl hunters harvested 5.8 million mallards per year. Since then harvest has fallen and about 3.5 million birds were killed annually during 2013 2015, for example. Of course, harvest levels vary across the range and depend on many factors. One is number of hunters, which during 2021-2021 was about 1 million. This resulted in a total duck take of about 11 million in the lower 48 states (according to Delta Waterfowl). As has been widely appreciated, the number of hunters has declined markedly since 1970, when it was estimated that 2 million hunters set out to put duck on the table.

The popular greenhead continues to be the most harvested of our native ducks, with roughly 2.8 million taken in 2020 2021. Certainly, I rate them high on the list of what I'd like for dinner (assuming I have no canvasbacks, which is almost a given). For comparison, hunters harvested 1.4 million green winged teal, 1.2 million gadwalls, and 1.1 million each of blue winged/cinnamon teal and wood ducks.

So, one might ask if a mallard is a mallard is a mallard? Short answer, no. In the scientific journal PlosOne, Michael Schummer and colleagues published a paper in March 2023 entitled "Population genetics and geographic origins of mallards harvested in northwestern Ohio". Their goal was to determine the genetic ancestry and geographic origins of mallards harvested in NW Ohio, to see how many were wild, game farm derived, or a mixture. To do this, they sampled 296 hatch year mallards that were shot by hunters. You might think of them as a waterfowl genetic ancestry program. Additionally, by analyzing stable isotopes extracted from feathers, they were able to estimate where the birds came from.

The authors noted that mallards and people have been closely linked, since the ducks were presumably domesticated in China around 500 B.C. Most domestic ducks are likely descendants of wild, Eurasian mallards, even the white Pekin duck we often see in parks. Supposedly, these were derived from 4 birds that survived a voyage in the early 1870's from China to Connecticut.

We also know that mallards have been extensively domesticated. I took part in one European tower hunt many years ago in which the targets were pheasants and mallards, both captive raised birds. Because of costs, tower hunts at least in my areas, don't include mallards, instead pigeons and pheasants are used (again, two introduced species). Now if they released Eurasian Collared Doves, I'd be out in a flash, as these are prime table fare (and no season or limits).

The point of that digression was that some of these captive reared mallards escaped. Many captive reared mallards in fact are released, often deliberately, and become part of "wild" populations. Just like I suspect that any pheasant has a captive reared bird within a few generations back in its pedigree, same is probably true for mallards. By the way, it is thought that escaped female ring necked pheasants are almost never successful breeders in the wild, but no one is sure about the reproductive success of roosters.

The magnitude of captive mallard releases is high, with 500,000 birds released annually from 1920 to 1950 along the Atlantic coast, and since them about 200,000. In other areas, releases were of lesser magnitude, but they were large, nonetheless. Incidentally, the USFWS estimates that there were between 7.5 and 10 million mallards in the US in 2022.

Schummer and colleagues used some modern methods to answer their questions. With modern genetic techniques, available in the last decade or so, it is possible to tell wild mallards apart from captive reared ones. They analyzed thousands of genetic loci and were able to assign the genetic ancestry of each harvested mallard to these categories: wild, game farm, and classes of game farm X wild hybrids. I'm not sure what waterfowl biologists think about the breeding fate of captive released hens.

The genetic results were, to me at least, fascinating. Of the harvested birds, 35% were pure wild birds, 12% were early generation hybrids between wild and game farm birds, and the other 53% were hybrids of a more advanced kind. That is, many generations of hybridization back in time.

These results raise several interesting points. One is that captive reared mallards and wild mallards easily fall in love, as can be seen in the image from Lavretsky et al.'s study. I mean in theory, we have gone from 100% wild to 35% wild, so there have been a lot of mixers between the two camps. Another point is that it might be impossible to tell from the plumage whether the bird was captive or wild or a mix. The latter observation is likely due to the fact that breeders of mallards want to keep the birds looking like mallards and not Pekin ducks.

My expertise is far from analysis of stable isotopes. So here is what the authors explained: "We also assigned these mallards to geographic natal origins using the ratio of stable isotopes of hydrogen (2H/1H) from feathers (hereafter δ2Hf) grown on natal areas because these isotopes produce a strong, predictable latitudinal gradient across broad geographic regions". Ok, in other words, these isotope ratios are like bar codes that identify where the birds was hatched. This method is better than banding data, which tends to be idiosyncratic (ok, I love that word, you can substitute "not as good"), although together they improve knowledge of where mallards moving during the seasons.

The isotope results were also fascinating, especially when combined with the genetics. Genetically pure wild mallards came from parts of the range farther to the north and west than mallards genetically identified as hybrids. At a finer level, they found that 17% of all Ohio samples had isotype ratios that likely meant the birds came from western prairies, parkland or boreal regions of the mid-continent.

Examples of ducks derived from a mallard stock through captive breeding. From Lavretsky et al. 2023, communications biology (https://doi.org/10.1038/s42003-023-05170-w)

So what? Is this just an academic exercise? The real concern is that integration of captive genes into wild populations can be maladaptive. The reason is that although a captive mallard looks like a mallard, by being raised in captivity, it allows some genes that would otherwise be maladaptive to be retained and then transferred to wild populations. Taking their results to a broader geographic level, over 90% of mallards in Atlantic coast states and 40% in mid-continent areas have substantial game farm derived ancestry. In western states, less than 5% of birds had significant captive ancestry. There was also considerable variation in other areas (e.g., mid-continent southern populations) suggesting continued monitoring.

The question is whether any maladaptive genes deriving from captive ancestry will result in declines of mallard populations. The authors posed the question of whether declines in mallard populations are a result of infusion of maladapted genes from captive populations. For example, if the migratory urge is bred out of captive mallards, these genes responsible that are introduced into wild populations might negatively influence wild populations that come from areas where migration is beneficial.

Few hunters would likely heel under a rooster they thought might be from a game farm. It might be hard to reverse the changing genetics of mallards, but maybe the goal is to have lots of mallards. Whether a dawn sky full of mallards of captive ancestry setting wings and coming into our decoys is an issue will be a question for the future. However, if a sky full of captive bred mallards comes at a cost to the genetics and future adaptability of wild living mallards, it will be a colossal management issue.

67. How's Your Kooikerhondje?

Cabin Talk.—On a trip to coastal town of Gaast in the Netherlands, my host asked if I'd like to see a "duck decoy." I figured something was lost in translation, so to be polite I said "sure." There's more to duck decoy than I thought. It took me on a trip back in time.

This flat lowland area was originally treeless. Ducks frequented the area in winter and were fairly common but spread out. The ingenious idea of how to concentrate and capture lots of ducks in a short time originated in this area by at least the 17th century, and it's now called a duck "decoy", which comes from the Dutch word "Kooi" for cage or trap. Here's how it worked.

First, a pond is surrounded by planted trees, to give passing ducks the illusion of a safe haven in an otherwise treeless landscape. The total size of the pond and the trees is only about 150 by 250 yds. On different sides of the pond, small channels about 3 yds across lead into

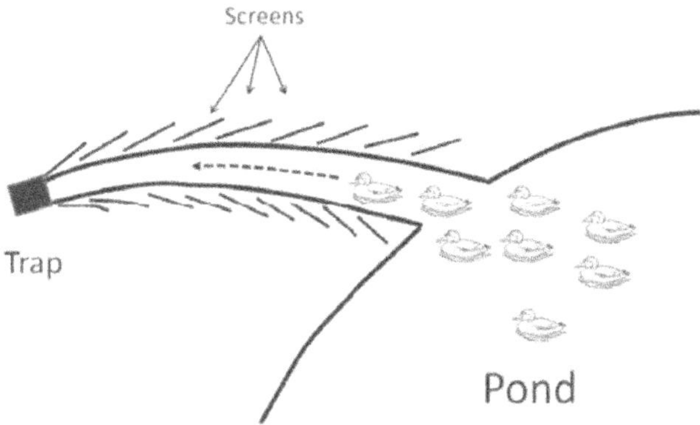

Schematic drawing of one of the channels leading away from the main pond

the trees. This particular duck trap had six channels leading away from the center pond like spokes. Covering each channel is netting raised up on poles, and along the channels is a series of 15 foot long matt walls or partitions, each about six feet high, which are arranged almost parallel along the channel, but at sharp angles (see diagram). It is possible to walk behind the partitions towards the endpoint, without being seen. The channels have at least one bend or curve leading to the endpoint; the point of each will soon become obvious.

The keepers raised ducks, called decoy ducks (often very noisy), so each pond had a supply of live "decoys." When the migrants visited this lowland area in fall and winter, they were naturally attracted to the tree surrounded ponds and the live decoys. Ok, so a bunch of wild ducks land in this pond. Did the locals open up and start shooting? Nope, these guys were doing this before there were decent guns. Plus, a few shots and it would be over.

How do you catch a bunch of wild ducks on the pond if you're not going to shoot them? The keeper lured the decoy ducks deeper and deeper into the channels by staying just ahead of them, out of sight, and throwing wheat or other grains on the water's surface. The wild ducks didn't seem to mind being invited to a free lunch, and what the heck, their new friends on the pond weren't afraid. I was thinking as this was explained, this is a good trick so far. You get a bunch of wild ducks heading down a covered channel – what next, get a net?

No, my host told me that you use your "Kooikerhondje." Ok, I'm lost again, more translation problems? Not sure if I've got one, or if I do, whether it still works! The Kooikerhondje, I learned, is a small spaniel like breed whose job it was to stay mostly hidden behind the screens alongside the channel and help lure the ducks down the channel and under the netting. Now, I'm really curious, not to mention skeptical. How does a dog "lure" ducks? The Kooikerhondje has a really big and fluffy tail plume (like my English setter) and the ducks could occasionally see it as the dog went silently back and forth behind the narrowing openings between the partitions, following hand signals from the keeper. The ducks, believe it or not, are curious of the dog's tail plume, and keep following it down the channel away from the pond. When the ducks are far enough down the channel, and around the bend and out of site of the main pond, the handler appears.

That sounded a bit counterproductive to me because for my nickel, the ducks, especially the wild ones, would exit to the main pond at warp speed – game over. Turns out the decoy ducks were annoyed but just swam back out into the pond, as I predicted. The wild ones,

Kooikerhondje, image by Shgadipositas, Pixabay

however, don't want any part of the Kooikerhondje-keeper action. Because they can't see the main pond, having gone around the bend, they fly/swim as fast as possible down the narrowing channel, towards the light (and sometimes a mirror) at the end (see sketch). For the wild ducks, the journey ends at the "catcher," a box where they are captured and killed. The decoy ducks are now back in the main pond having had a nice lunch and are ready for the next unsuspecting group of wild ducks to appear.

The use of these duck decoys (as the locals call the whole operation) is now illegal for hunting. But some of the sites are still used for catching ducks to be banded for tracking and research, because it's very effective. Some even still use the Kooikerhondje. Seeing it first hand was a trip into the history of duck harvest I didn't know existed. There are lots of old duck decoys, and some like the one I visited near Gaast are maintained by dedicated volunteers, as a historical site. If you get a chance to visit this region, try to arrange a visit. You'll find it a fascinating experience. But remember, this area was enriched by lots of duck droppings and I've never seen a bigger, taller or denser patch of stinging nettles! Maybe it's the revenge of the ducks.

68. Walleye And Perch, Same Family, Distant Relatives, Same Taste?

Cabin Talk.—Although my professional career has involved figuring out how populations and species are evolutionarily related to one another, I never stopped to think about perch and walleye. Other than the countless times I've discussed with someone about how I love catching perch because they taste like walleyes -- after all, they're in the same family. But does being in the same family mean they are closely related? It dawned on me that this has been studied, and sure enough, there are a bunch of papers showing how they're related. In sum, they're distant relatives at best, in a taxonomic family with almost 250 species.

For starters, there are five species in the taxonomic genus *Sander* that includes walleye and sauger from North America, and three European species: pikeperch (image), estuarine perch and Volga pikeperch. A genus is a taxonomic rank that indicates that all of the included species are more closely related evolutionarily to each other than any of them are to species in any other genus. That is, if you told me that you discovered a new species in the genus *Sander* I could predict many of its characteristics. The taxonomic framework, tracing to the Swedish botanist Linnaeus, is predictive.

The yellow perch we all know, and at least I love, is one of three species in the genus *Perca*. Our yellow perch is most closely related to Balkhash perch, which

PikePerch from Pixabay (Andreas Barsch)

lives in mostly in Kazakhstan! How that could be is another story. From images online, the Balkhash perch looks really similar to yellow perch. The third perch is the European perch, which lives from western Europe all away across to eastern Siberia. You'd recognize them immediately as a perch as well. European perch have been introduced into Australia and have caused considerable damage to native species; they are considered a noxious species in New South Wales.

So far, it is not obvious how closely related yellow perch are to walleye – are walleyes and perch "sister" groups, or each other's nearest living relatives? Far from

Balon's ruffe (one of 4 species of ruffe) occurs in parts of Eurasia (image from Europe Zoo). It reaches a length of 6 inches. In spite of appearances, Balon's ruffe is more close-ly related to walleye than it is to yellow perch. Walleye and perch at bottom from northern MN.

it. To illustrate how closely related they are evolutionary biologists would show you a phylogenetic or evolutionary tree with 250 species, which would be hard to read unless you were interested in the nitty gritty details.

Let me put it this way. Walleyes are more closely related to a couple hundred other species including darters in the genera *Etheostoma*, *Percina*, *Ammocrupta*, *Crystallaria*, *Zingle*, and the ruffes, than they are to yellow perch. The accompanying image provides some insight. The Balon's Ruffe is more closely related to walleye than the ruffe is to yellow perch. The fact that the ruffe and the perch look similar is superficial and not indicative of close relationship. We would say it's a "convergent" appearance, hit upon independently as a response to either predation or competition.

It is true that people who study fishes place perch and walleye in the same diverse taxonomic family. In that sense they're "closely related," at least compared to fishes in other taxonomic families like groupers and sea bass. Of course, in our case, the real test would be a blind taste test. You'd predict from the taxonomy that if perch and walleye taste similar, so would all the other species! That would mean eating a lot of darters (beware many are threatened or endangered). I've often wondered after an unproductive day on the lake what our remaining shiner minnows would taste like. If you know please tell me.

69. Talk About Opening A Can Of Worms

Cabin Talk.—My wife and I once had some cod, and I noticed a white filamentous thread that I figured was some kind of connective tissue. After eating a couple of these, I looked more closely and realized that they were nematodes, a parasitic worm. My wife stopped eating. When I googled them, I found, unsurprisingly, they're called "cod worms." If cooked, they only add to your meal.

Salmon also have nematodes called anisakids (a similar worm as in cod), which are not considered particularly detrimental to the fish itself. The life cycle goes like this: marine mammals ("definitive host") shed anisakid eggs in their feces, eggs hatch and the new larvae are in the water column until they are ingested by crustaceans like krill (shrimp-like) although the larvae survive, salmon ("intermediate host") eat the infected crustaceans where the worms develop further, and seals and whales eat infected salmon, where the parasite matures into an adult worm in the mammal's gut. If you're on the parasite's side, mammals and salmon are a parasites way of making another parasite.

Owing to the recovery of populations of several marine mammal species, the potential for these parasites to increase is heightened owing to a greater abundance of hosts. Because so many animals depend on salmon, not to mention humans, monitoring the parasite population becomes important in several ways. But how could one figure out whether the frequency of parasitism had increased?

There are too few specimens of salmon in natural history collections, and records kept during salmon processing aren't very systematic. By chance, Dr. Natalie Mastick from the University of Washington, Seattle, and colleagues discovered an industry organization (Seafood Products Association) that has preserved cans of salmon for over 40 years, to determine the integrity of the fish over long periods. The samples from Alaska included salmon identified as chum, coho, pink and sockeye. Personally, although I routinely keep cans past their expiration, I would be disinclined to consume 40 yr old canned salmon. However, Mastick and colleagues realized these cans of decades old salmon could provide a historical record of levels of parasitism in salmon. That is, the worms were canned right along with the salmon, although not for the purpose Mastick put them to!

The team was able to analyze cans of 62 pink, 52 sockeye, 42 chum and 22 coho spread from 1979 to 2021. Unsurprisingly to me, and vindication for my excluding them from appetizers, some of the fillets had basically become one with the liquid in the cans. They also realized that a single can might include more than one fish, and a large fish might have been used to fill multiple cans. But there are issues in all data sets.

So, what's involved? The meat was dissected into small portions, and by looking for capsules or pockets in the flesh, produced because the worms form a coil (see image), they were able to find anisakids, which like good scientists they preserved in 70% ethanol. They noted that even though the older cans were degraded, the worms had maintained enough integrity to collect and identify. In case for some reason you were wondering, anisakids have a larval tooth and a ventral excretory pore that looks like a dimple. However, I find nothing cute about the ones I ate in cod. Canning kills the parasite, as does freezing and cooking.

What did they find? Out of 178 cans, 50.5% contained nematodes, including 57.1% of chum, 27.3% of coho, 46.8% of pink and 59.6% of sockeye. These represented a total of 372 nematodes. However, in one can of chum from 2019, there were 155 worms! It makes me wonder if whomever was doing the canning wasn't in fact asleep at the proverbial wheel.

And the results are in: "We found a significant increase in anisakid burden in chum and pink salmon over the 42-year study period. We found no significant change in the number of anisakids in coho and sockeye." The authors suggested that the increase in anisakid abundance was due to population increases of northern fur seals, harbor seal, humpback whales, belugas, and several populations of killer whales.

In addition, the authors noted that the increase in sea surface temperatures over this period could affect parasite development because the anisakid eggs develop faster at higher temperatures, during their brief time when they are free-living in the marine environment (otherwise they're buffered by their host tissue temperature).

A coiled anisakid from "Opening a can of worms: Archived canned fish fillets reveal 40 years of change in parasite burden for four Alaskan salmon species" by N. Mastick and colleagues (2024, Ecology and Evolution).

We might ask who cares if there are more or less worms in salmon. If anisakids are consumed alive (undercooked or raw salmon) it can cause anisakidosis, or food poisoning. Is there any good news here? Yes, increasing levels of parasite infection suggest that the ecosystem is returning to a healthier state because of the recovery of the mammalian host species. In fact, if not oddly, we might think of the level of anisakid parasitism as a proxy for ecosystem health.

The most common surface for an animal to live on
is another animal - the world is a parasitic one. This study
used a novel source of data, canned salmon, to monitor
the levels of nematode infection over time. Paleontolo-
gists dealing with small bone fragments will be jealous -
digging in rocks under the blistering August sun doesn't
compare to opening a can in the lab. I can only tip my hat
to the authors from coming up with an amazing way of
monitoring populations over several decades - a can open-
er and a forceps.

70. It's Heating Up Beneath The Waves
Too

Cabin Talk.—Few days go by without media attention to global warming. Writing about global warming
is actually pretty low hanging fruit because the evidence is
unambiguous: the earth is warming. This is nothing new,
at least when we recall that 21,000 years ago there was a
mile thick glacier covering parts of the upper Midwest, one
of over 20 such glacial advances in the last 2 million years.
All glacial retreats occurred because of global warming.
However, the new part to the current warming trend is
that climate (not weather) scientists think the warming is
more rapid than in the past, and that there's a strong like-
lihood humans have something to do with it. The effects
of a too-rapidly-warming earth are seen just about everywhere, from melting glaciers to early nesting birds to more
frequent weather "events."

In spite of all I have read and studied about global
warming, and in spite of the fact that I spend a decent
amount of time fishing, my attention to water temperatures usually is a yearly one. Finding crappies and walleyes

can be water-temperature dependent, which of course changes over the open water period. For example, on Minnesota lakes I fished in 2025, water temperature was 61° on May 27 and 75° on July 3, and 67° on Sept 25, with variation between days and lakes.

But what about longer term trends? Is the well-established global warming in earth's air temperatures also "echoed" in the water? Minnesota DNR published average lake water temperature from 1968 to 2022 for the month of July across Minnesota lakes. I contrasted those data with air temperatures (maximum, minimum, average) from the same month and time period. I was interested in whether the air-water temperature trends would be similar.

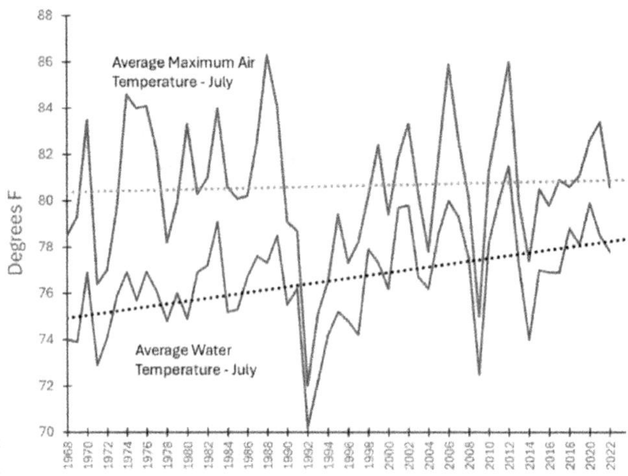

Image 1. Average maximum July air temperature and water temperature averaged across Minnesota lakes.

As you might expect, there's a pretty strong relationship between air and water temperatures (Image 1). When air temps are higher or lower, so are water temps. However, there are some peaks in maximum air temperatures in the 1972-1974 and 1986-1990 periods that are not accompanied by major spikes in average water temperatures. The 1992 and 2009 drops in maximum air temperature did correspond to a lower average water temperature.

July water temperatures are likely related to those earlier in the year, so there are many confounding variables.

The plots were consistent with the general pattern in global warming where minimum temperatures increase more than maximum temperatures. For example, mean January air temperature in Minnesota went from -2.2° in 1968 to 4° in 2022. In contrast, the average maximum July air temperature only increased from 77.2° to 77.9°. If the

Image 2. Eurasian watermilfoil on shore of Leech Lake, Minnesota, June 28, 2025

minimum temperature increases but the maximum stays about the same, the average temperature goes up.

Most interesting to me is that the average water temperatures increased by several degrees. Now, I realize this is just July, and it's an average across Minnesota lakes, but an increase of 2 to 2.5 degrees over about 50 years is actually a lot. When water temperatures rise, it affects many aspects of aquatic life. Increased algal blooms are one negative impact. I don't know if it's related, but the abundant floating mats of Eurasian watermilfoil on Leech Lake so far this summer, like the one at my shoreline (Image 2), are impressive but depressing. Maybe warmer water enhances Eurasian watermilfoil growth. Increased water clarity due to zebra mussels might aid milfoil growth because light reaches greater depths. It would be really nice if zebes ate milfoil.

Speaking of zebra mussels and water clarity, according to MN DNR after Zebes were observed in Winnibigoshish during the period 2012-2016, lake clarity increased from an average of 6-7 feet to readings of from 9 to 27 feet in 2024. MN DNR notes that increasing water clarity tends to improve habitat for bass, sunfish, and

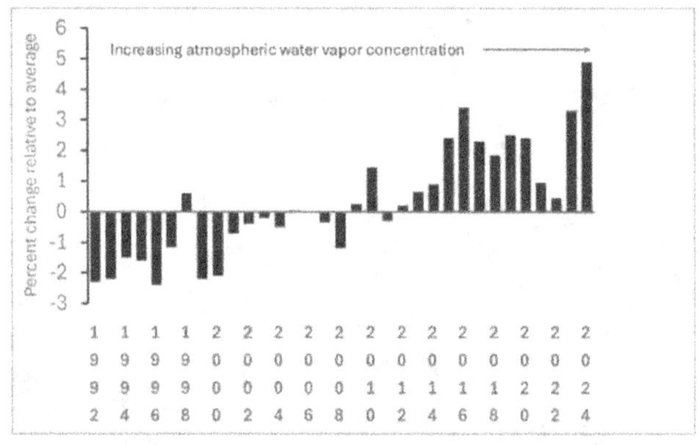

Image 3. Annual values for water vapor concentration from 1992 to 2024, showing generally increasing values starting about 2014.

northern pike but decreases habitat suitability for Walleye – the darker depths used by walleye for foraging are getting deeper, and if the lake is too shallow, they might be squeezed down and out so to speak.

Water temperature affects where insects and other aquatic invertebrates (read fish food) will be hanging out.

And, increasing water temperature affects dissolved oxygen, which we know influences where fish and their food are found. Warmer water has a lot of potential impacts on our fishing environment. Some broader scale climate shifts are also likely to influence our lakes, forests and grasslands into the future.

Increasing (record) numbers of flash floods are predicted with global warming, because warming temperatures increase evaporation, which increases water vapor in the atmosphere, resulting in heavier rain storm events. On the other hand, in some areas increased temperatures and evaporation lead to droughts. In addition, long-lasting droughts reduce plant cover and make flash floods even more devastating when they do occur.   As climate (not weather) scientists have been predicting, global warming leads to increased weather variability – heavier rains, more frequent droughts and more severe weather events.

I don't have a background in climate science (I'm a bird guy). To determine if what I read was true, I found an informative visual of the change in the amount of water vapor in the atmosphere produced by Copernicus Climate Change Service (Image 3). Earth reached a record value for total amount of water in the atmosphere in 2024, at 4.9%. So, recent flooding is not surprising.

We're in a time in which the environment is changing rapidly, both on land and in the water. We have a decent grasp of the effects on land-dwelling creatures but what it holds for fish and other aquatic organisms remains to be seen. Anything we can do to stall the current warming trend can't be a bad idea.

DOGS, BUGS AND
STUFF

71. How Long Have Man And Dog Been Best Friends, And Where Did It Begin?

Cabin Talk.—Twenty thousand years ago, you're living in small groups in caves or makeshift shelters south of the great glacier that extends to present day Nebraska. You share the landscape with lion, cave bear, horse (not the reintroduced one), camel, mastodon, woolly mammoth, sabre-toothed cat, American cheetah, a beaver weighing 150 lb, two species of tapir and a large ground sloth. Some of these animals are great eating, others would prefer to eat you. Your weaponry is primitive, and your senses need to be keenly tuned to the presence of any large predators in the area. You use fire to keep warm, cook your food and keep the dangerous predators at bay. Packs of wolves wander the countryside, and you compete with them and other predators for game and left overs from kills by the big predators. It might have gone something like this.

Around your camp during good times there are a few scraps of food or bones. A lone female wolf, injured, limping and evicted from her pack, sneaks in to gnaw on a bone or piece of gristle. You're well fed and the wolf doesn't pose a threat. Maybe you toss it a scrap, and over days or weeks, it starts to lose its fear of you, and vice versa. Perhaps the cave man in you lets down, and you actually feel a little compassion and even feed it once in a while. Seeing your competitor in a new light is intriguing. You're both social hunters after all.

One warm afternoon, the wolf is back, lying at a distance, and you're all taking a siesta. A warning growl from the wolf brings you to your feet just in time to grab your spear and kill an attacking lion that intended to grab lunch, you. A primitive mutualism is born. You feed wolf, wolf provides even keener eyes, ears and nose. Protection for you, food and possibly protection for the wolf. It's a win win situation, maybe a bit tense at first. Later in the spring, the female gives birth to a litter, and the pups gradually become part of your clan. Watching them play makes you smile, and the women swoon. Loving puppies is probably a deep seated emotion.

However it got started, humans began to domesticate dogs from wolves a long time ago. In fact, it is thought that dogs were the first domesticated animals, before kittens and chickens, cows, goats, sheep and pigs. The result is an astounding variety of dog shapes and sizes, from Great Danes to Chihuahuas, guard dogs to lap dogs, hunters to show dogs. All of them, we think, produced by selective or artificial breeding from ancestral wolf stock. Wolves on the other hand, have continued to look pretty much like wolves for a long time.

Resolving the time and place of dog domestication can be approached by studying fossils and DNA. Fossils provide hints as to where and when dogs were first domesticated because bones (especially skulls) of early dogs differ from wolves. If you find bones from both early dogs and wolves at the same place, you have a clue as to where and when domestication was in progress. The oldest known dog like fossils are from Western Europe and Siberia, and date from 15,000 to 36,000 years ago. Some scientists had proposed that dog domestication began in the Middle East, Eastern Asia, or both. However, the oldest known dog fossils from the Middle East or Eastern Asia are less than 13,000 years old, but fossils are notoriously rare and no one puts much stock in their absence. At present, the fossils favor earliest domestication in Europe.

Comparisons of the molecules of heredity, DNA, can also help identify the time and place of domestication. Analysis of variations in DNA provides a genealogy, in fact, an extended genealogy. At the level of an individual, its DNA profile is often a bar code of where it is from. At the level of species, DNA profiles reveal how species are related (with some real surprises, for example, that whales are actually closer to hippos than anything else!).

Today scientists do not need to rely on fossils or DNA separately, because sophisticated modern techniques permit DNA to be recovered from fossils of known age. A paper in the journal Science by O. Thalmann and 30 (!) other colleagues compared DNA from 18 fossil dogs, 77 modern dogs, 49 modern wolves and 4 coyotes. Their findings suggest a number of interesting insights into man's domestication of his best friend.

The dog/wolf genealogy shows that there are four major groups of dogs, and their family histories are intermingled with those of wolves. In other words, not all living dogs are most closely related to other living dogs, and not all wolves are most closely related to other wolves. This finding means that dogs and wolves have separated very recently, because distinct species have diagnostic DNA profiles, but this takes (evolutionary) time. The structure of the dog/wolf genealogy also means that whether you consider dogs and wolves to be separate species depends on what you think a species is! One could argue that they are separate species that can still interbreed, at least some of them. Others would argue that they are the same species. Irrespective of whether dogs are species, the genealogy clearly supports the notion that dogs came from a wolf ancestry.

What was relevant to determining where domestication first began was the observation that the four distinct groups of dog DNA are more closely related to fossil dogs from Europe, not wolves from the Middle East or Siberia. It is possible that dogs were also domesticated in this region much later than those in Europe, but the fossils support the initial domestication in Europe. The next question is when did it occur?

Based on the ages of fossil dogs, and the DNA similarities of fossil and modern dogs, it appears that domestication in Europe occurred between 19,000 and 32,000 years ago. This is a frustratingly large range of dates, but it's the nature of the beast. It at least sets a time frame that excludes very recent times. Given that modern humans originated in Africa around 150,000 years ago, it means we only started domesticating wolves very recently in our history.

Another conclusion that followed from the DNA genealogy was that dogs likely arrived with the first humans in the New World, crossing the Bering land bridge from northern Asia. This might suggest domestication closer towards 30,000 years ago in Europe, giving people and their dogs time to spread across Eurasia (and be found in the Middle East). Also, domestication of dogs preceded the development of agriculture. This observation might shed light on what the earliest domestic dogs looked like – obviously not Chihuahuas but working dogs that would aid hunter gatherers.

Bottom line: when you sit down to have that talk with your dog about where he's from, you can tell him the story about how his early ancestors were wolves that formed bonds with humans who were excellent and fearless hunters, much like yourself, at least 19,000 years ago in Europe, and that they walked onto this continent via the Bering land bridge. If you want, you can embellish it with how the early dogs never had an accident in the cave, always obeyed, retrieved soft mouthed to hand, never strayed into the neighbor's yard, and had absolutely no idea what happened to the cat.

As newer and newer genetic techniques emerge, our understanding of dog wolf relationships continue to change, at least somewhat. An archaeologist and geneticist Greger Larson from Oxford University (London) has been gathering DNA from ancient dog bones, and was given a DNA rich bone (the petrous bone, from the ear) from beneath a 4,800 year old monument at Newgrange (which predates the pyramids of Giza and Stonehenge).

When Larson and colleagues put the new DNA sequences together, they discovered that there was a major fork in the dog lineage, one including dogs from eastern Eurasia (including Shar Peis and Tibetan mastifs) and the other included all western dogs, and the one from Newgrange. The emerging idea is that dogs were originally domesticated in what is modern day China and were taken west.

There is, however, considerable disagreement in the scientific community, apart from the dual track of domestication and a southeastern Asia origin event. Not everyone agrees, but dogs were probably domesticated at least twice from gray wolf stock. They were moved about, and one of the lineages might now be extinct, with dogs from the eastern stock now worldwide. In any event, much research is ongoing so the story is probably not in final form. But that's the nature of science — new evidence, new interpretations. You have to be prepared to abandon your once cherished ideas in the face of new evidence. Given how many dog breeds there are, it's perhaps not surprising that figuring out ancestry is complicated.

72. Keeping Old Dogs In The Loop

Cabin Talk.—All dog lovers have heartwarming stories about their dogs, all of whom attained exalted family status. Upon becoming empty nesters, our dogs became our kids. When you have to say goodbye to a family dog, the grief lingers for a long time, although I feel we have given each of our dogs a great life. Sometimes that can come in odd ways.

Our dog, a Drahthaar named Zeke, developed an elbow infection a few years back that cost a couple of

Zeke, the old man in back, and Nuke; both are Deutsch Drahthaars.

thousand bucks to fix. When I thought he was healed, I took him out for a few minutes, and it was clear it had not gone away. So, he became a valued family house dog, retired from hunting but with many great memories.

He knows full well when I'm taking the younger dog, Nuke, on a hunt, and he's out at the back of the truck hoping for a kennel spot, so it's really tough to have to put him back in the house and leave him behind. It was way harder at first, but at 14.5 yrs old, a treat distracts him long enough for me to get away.

Still, it's a drag to leave him out, but as I too age, I'm starting to realize that I will appreciate a bone or two every now and then. So, to keep Zeke in the loop, when we come back with some birds, I line them up in the garage and send Zeke from my cleaning station in the house to "fetch" them, and he dutifully brings them back one by one (Figure 32). He even growls at the younger one who is also trying to get a bird (that he already retrieved in the field), reminding the younger one that he's still dominant, even though he's not capable of backing it up anymore. For Zeke, tail wagging mightily and with an extremely pleased, if not proud, look on his face, it's a bit of redemption. If you have an old dog, give it a try. And maybe when we're old and out of the field, someone will let us make a garage retrieve or two. Even a small redemption is worth it.

73. Dishonest Doggies

Cabin Talk.—We humans sense a small part of the world's offerings. The scents or pheromones that female moths give off to attract males is lost on me. I know because of discoveries a couple of decades ago, that many animals can see and communicate via the UV part of the electromagnetic spectrum (between visible light and x rays), but that too goes right by me. Some other things happen with my dogs right under my nose, and yours as well.

Humans carved dogs from wolves using a lot of selective breeding. From Chihuahuas to great Danes, the original wolf signature is pretty much obscured. Given all that we made out of dogs, it might come as a surprise that there are a few things that we don't yet know.

For example, when a dog comes out of the water it shakes its fur almost violently, spraying water, and whatever is in it, in all directions, mostly mine it seems. A scientific study showed that dogs start shaking from the head, and because of having very loose skin, dogs generate maximal centripetal force that dislodges 50% of the water in less than a second. If you're longing for physics class, it is actually breaking the water's surface tension, so it flies off the fur. Incidentally, if you were thinking of saving towels and wanted to try this yourself after your next shower, forget it. Your skin isn't loose enough, and all you'll do is make your chiropractor richer.

In calm periods of the earth's electromagnetic field, which isn't all of the time, dogs orient their bodies in a north south geographic axis when doing their business. That is, they have a magnetic compass and they use it to sense direction. Why they do this is a tougher issue, but possibly, if you're an ancestral wolf with a compass, you can mark the boundaries of your territory with accuracy. Domesticated dogs haven't lost the skill.

What about some other things we might not have thought of? Consider male dogs raising their legs to urinate. I admit to thinking about it before and noticed that this behavior helps aim the urine away from the backs of the front legs. Imagine you're a scientist on a slow day. You're thinking about male dogs and how they raise their legs to urinate. Your family has, for some unknown reason, a small dog, say a pug. It too raises its leg to pee, but you casually notice that it's at such a steep angle it has to catch itself from tipping over, and you wonder what's up with that. Is it just the effects of rampant inbreeding?

Male mammals have at least two reasons for their peeing posture, one is obvious, to pee. The other is as mentioned above to mark their territories. Dog urine contains messages about including individual identity, sex, age, reproductive status, social status, health, quality, kinship, and histocompatibility. Whew, no wonder my dogs are so fascinated by sniffing where, I presume, another dog already peed. Is it a girl? Do I know her? Is she in heat? Can I take this guy? Is he feeling sick? Am I related to him? So many questions, so little urine. But, what about that pug and its leg angle?

There's a way to accent the messages - get your pee as high up as you can on your target, whether a bush, a fire hydrant, or in the case of my drahthaars, the car tires of visitors. A scientific paper by B. McGuire and colleagues (from Cornell University in Ithaca, NY) wrote an entire paper on urine marking by male domestic dogs. Like I said, must have been a slow day, but they made some interesting discoveries. Think of the physics. Urine that is higher up can be carried farther downfield by wind and get disentangled from ground scents. Also, it puts the urine at nose height, although I doubt dogs need any help finding it. Their slant was this: is it an honest or dishonest signal? In other words, can dogs lie with how high they pee?

Now, imagine you're a dog that comes along and sniffs this scenty brew. If it's pretty high up in the bush, doesn't that potentially say it was left by a relatively large dog? That's where McGuire started thinking about small dogs. Believe it or not, they set out to determine if the raised leg angle predicts urine height in the bush, and whether small dogs elevate their hind legs more to get the urine higher. Here's what they found: "small adult male dogs may place urine marks higher, relative to their own body size, than larger adult male dogs to exaggerate their competitive ability." They do it by raising their legs relatively higher than bigger dogs! In other words, the height of a small dog's urine might be a dishonest signal of his size! In essence, he is hoping that if they were still wolves, and this stuff mattered, a small stature individual could "lie" about his size and send a (dishonest) message to other wolves in the vicinity that they should watch out because he's big and nasty.

Reminds me of a saying, slightly edited: if you can't pee with the big dogs, don't raise your leg.

There are many things we have yet to learn about dogs and what they do. It never occurred to me that small dogs might try and pee a bit higher, and I understand why it might be adaptive in a wild canine. This clever study uncovered a holdover or legacy of an ancestral time when urine height mattered. Our domestic dogs are simply expressing a behavior they inherited from their wolf ancestors. Kind of like the 'appendix' of behaviors.

Another likely message contained in urine is what kind of dog it is, and there is research to back this up. I would wager that my dog knows a pug's urine when he smells it and other than being curious, wouldn't for a minute consider it threatening, even if it was "posted" at a height that seemed to indicate a dog twice pug size. Pugs pee pug pee. Still, the importance of recognizing the type of dog might be historical in nature. Once upon a time, say 12,000 years ago, gray wolves and the much larger dire wolves lived side-by-side. Knowing whether there was a dire wolf in the vicinity might have been pretty important to a gray wolf, and has been carried down to your dog, pug or drahthaar.

An interesting test would be to measure leg angle in wild canids of different size in the same species. I don't think that it would be informative to compare say a fox and a coyote – no doubt they can tell who left the urine, and a coyote is just not going to be afraid of a fox no matter how high he gets his urine. Actually, a fox doesn't want a coyote to know he's around because coyotes kill foxes! Still, it's interesting that these behaviors are still expressed in domestic dogs. Is it really adaptive, or do these behaviors still exist simply because enough time has not passed for the behaviors to disappear? I'm guessing the latter.

74. More On Your Dog's Potty Habits

Cabin Talk.—I chuckled when I started this section because I realized there was nothing I could say about dogs that hadn't been said before. Still, I'm fascinated by enduring questions. What are dogs thinking, or do they think anything like we do? Exactly how much to they understand when I talk to them, do they get innuendo and humor? Do they roll their eyes like my wife? How I wish they could tell me what they're smelling in the distance. I am fond of telling students that it's perhaps fortunate that humans have lost the keen sense of smell so we aren't constantly sniffing everyone's butts.

We do love man and woman's best friend, our dogs. But we often don't understand why they do what they do, and sometimes we don't even know that they're doing something that is obvious to them but not to us.

I've read that the tight circles that dogs make before laying down to sleep have several potential reasons. One, it allows them to be sure of the wind direction so they can sniff out any danger coming from that direction. Another is to make a comfier bed. Granted your dog doesn't have these concerns, but these are thought to be behaviors inherited from wolf ancestors who depended on them for survival. So our dogs do this routinely in our bed, several times a night.

An earlier study examined the direction dogs face when urinating or defecating. The authors of the study wondered if dogs used their ability to sense the magnetic field and oriented themselves in a northward direction when urinating or defecating. Actually, when the authors first analyzed their data, they found what I would have predicted – dogs, especially males, relieve themselves in random directions. However, the earth's magnetic field goes through "calm" periods and during these conditions dogs oriented their body axis in a N-S direction when defecating and urinating. Why? That's a good question.

Another study found that small male dogs raise their hind leg relatively higher when they urinate. They reasoned that getting the urine higher in the vegetation or on a tree would send a false signal about the dogs' size to another dog who happened upon the spot. Apparently bigger dogs, like my drahthaars, aren't fussy, because mine are just as happy to pee on the top or bottom of your car's tires.

In case you thought you'd heard it all, hold on. The McGuire group published a scientific paper last year that provides some intriguing insights into whether a dog reacts differently to an unfamiliar man or woman.

McGuire's group reviewed studies that show that in animal shelters, dogs react differently to men and women. For example, dogs in a shelter barked longer at an unfamiliar man standing in front of their cage than for a woman. Male dogs in a commercial kennel spent more time near an unfamiliar woman than near an unfamiliar man, whereas there was no difference shown by female dogs. It doesn't surprise me that I would be more threatening to a strange dog than my wife, but I'm not sure I can explain why. Perhaps it's because I'm bigger or my testosterone tainted scent might be a warning signal that I'm potentially more dangerous.

Under normal conditions, adult male dogs use the raised leg urinary posture about 95% of the time and adult females 19 37%. However, in "fearful situations" adult male dogs revert to the juvenile lean forward posture (all four feet on the ground). The behavior of dogs in shelters has been studied. Shelters try and get dogs out on leash to give them exercise and some social interactions. Both male and female shelters dogs use the raised leg urinary posture less frequently (73% of the time for males and 6% for females) than reported above for dogs under normal conditions. Stress of being at a shelter makes some dogs more cautious or fearful.

McGuire and colleagues noticed that there was no information on whether a shelter dog's behavior differed when it was being walked on a leash by an unfamiliar man or woman. Ok, maybe this is not the first thing you wonder about when you wake up in the morning, but we often learn from what we might initially consider far-fetched situations. Their idea was to see if the scent marking behavior of shelter dogs differed when being walked on leash by unfamiliar men or women. For example, they thought that dogs being walked by an unfamiliar man would more often revert to juvenile squatting behavior when urinating.

The study was very detailed, and they accounted for every possible confounding factor they could think of. The dogs were mostly mixed breeds, over one year old, almost all were spayed or neutered, and in good health. People assisting in the study chose dogs they had not walked before. Dogs were not allowed to interact with other dogs and the walkers recorded each urination and the posture used (squat, raised leg), defecations, and bouts of ground scratching. No, the field work was not glamorous.

The results were, however, very interesting. Male dogs walked by men urinated ½ as often as male dogs walked by women. Male dogs were more likely to defecate when walked by women (62% of walks) than by men (44%). Female dogs were less likely to urinate when walked by men than women, although the differences were not as big. Female dogs were more likely to defecate when walked by women (73% of walks) than by men (49%).

When male dogs being walked by men urinated, 21% used the lean forward all feet on the ground posture, whereas the figure was 13% when walked by women. That is, male dogs often reverted to a submissive behavior when walked by an unfamiliar man. Female dogs tended to use the squat posture whether being walked by men or women.

McGuire and colleagues concentrated on the relevance of their results for shelters. In particular, they noted that a dog's initial evaluation might be biased when evaluated by men, and some dogs never seemed to "warm up" to unfamiliar males.

Some behaviors we see in domesticated dogs are evolutionary holdovers from wolves. During the taming and domesticating of wolves, why would it have been adaptive for wolves to fear men more than women? Perhaps men were more aggressive. In the McGuire studies male dogs adopted more subordinate behaviors when confronted with unfamiliar male walkers. In my opinion, both urination and defecation might make a dog more vulnerable and therefore be less likely to do their business in the close presence of an unfamiliar human male. Of course, there comes a time in every dogs' life when if you gotta go, you gotta go, even if you're on a leash.

Maybe a fear of men has been reinforced since wolves were first domesticated. Studies show that men are more likely to be charged with animal cruelty than women, so perhaps wolves and dogs act accordingly. We obviously have a lot yet to learn about our best friends.

75. To Pee Or Not To Pee, That Is Your Dog's Question

Cabin Talk.—I gave some thought to an age old question that many, well maybe just me, have pondered. Let's say you're taking your unfixed male Deutsch

Drahthaar on a walk. The dog asks you how far are we going? You decline to acquiesce to his request – means no (apologies to Pirates of the Caribbean). Whether I'm going out on the road depends on the mosquitoes and deer flies. Might be back really quickly.

Why would your dog want to know how far you were planning on walking? Simple. If you're an unfixed male you need to be prepared to mark an unknown number of spots along route. What if you had just gone outside a short time before and your supply was running low? Your dog needs that information.

Obviously, if it was going to be a lengthy walk you have many spots needing a bit of pee, on both sides of your path. Your dog wants to be sure that any dogs walking the same track later know you were there first. If you knew upfront it was going to be a long walk, you engage in a bit of urine rationing. Or, as it seems with my dogs, you might hold out for spots other dogs had already marked, so you can at least top off their contributions. Now, if it's a short walk, you might want to hit a few spots a little harder, to make a point. So how long the walk will be can be useful information for your scent happy pooch's planning.

Scientists have actually studied quite a bit about dog pee and peeing. The hormones in dog pee are actually how dogs have a conversation with each other. A dog that is stressed out will inform the next dog in line by the stress hormones it its pee. Sexually mature males can send signals to female dogs about their being able and willing, or alternatively, these can be warnings to rival males that they should avoid crossing this way.

I was surprised to learn that males will sometimes raise their legs even when the tank is empty, and it becomes a visual display called a "raised leg display" or to me, more satisfying is the term "pseudo urination." If another dog is watching, apparently the intent is clear even if the scent is lacking. Or maybe they're just practicing.

My drahthaars are fond of peeing on the tires of every visiting vehicle, from cars to delivery trucks. The reason is, apparently, that tires pick up a lot of scents that tell a rich story to your dog about where that vehicle has been and what it has run over. Maybe the vehicle ran over a skunk, which would seem like perfume to my dog. So your dog is essentially mailing a message to the next tire sniffer down the road. Only one person was offended by my dog's affection for their tires, someone who came to pick up something I was giving away and he remarked in quite a huff that he was going straight to the car wash. Maybe he eats off his wheels, who knows?

I've never gotten down on my knees and sniffed my tires, but I'm pretty sure I wouldn't smell much other than rubber. But it's obvious that dogs get a good whiff of stuff. We think that a dog's sense of smell is 10,000 to 100,000 times better than a human, owing to the fact that dogs have a 50:1 advantage in number of scent receptors, and associated blood vessels, than we do. There are reports that some dogs can detect smells from 12 miles away. I have smelled plenty of smoke from Canadian wildfires from much farther away, but ok, I get it, it's not the same thing. And a dog can smell a substance at a concentration of 1 trillionth of a gram (0.000000000001)   for grins I tried converting it to pounds and the online calculator returned a value of 0!; I think I broke it.

I've often watched my dog, nose into the wind, sniffing with great interest and concentration, and thought to myself, "I sure wish he could tell me all the stories he was reading about in the wind."

76. What Might Mayflies Have In Common With Passenger Pigeons?

Cabin Talk.—Visiting a lake in May or June in much of the Midwest will likely bring you into contact with mayflies, although they also occur throughout the country. We sometimes see huge swarms in spring or early summer, and occasionally they show up on radar. It's common to find them attached to buildings or trees next to lakes after a hatch.

I found a bunch of mayflies attached to my place at Leech Lake in late June. Many were attached to screens and a smaller number directly to glass; in fact I counted 4.5 per screen and 1 per window glass. Sorry, sometimes I can't control my inner scientist. I thought it was interesting that they could attach to glass given the 15 mph NW wind that was blowing directly on the window. But then, I had one on my truck windshield a few days prior, and it held on to the glass until I reached 50 mph.

Wondering how they do this, I learned that mayflies could attach to glass with adhesive pads on their feet, or to screens with tarsal claws. I watched one move on my screen and it had to lift each foot up awkwardly to disengage the claws from the screen and set it down a short distance away. Tarsal claws obviously didn't evolve for hanging on screens! I admit to never having learned much about mayflies, other than anglers don't appreciate them when there's a hatch, and they sometimes make a mess on roads. I figured there was "a" mayfly. Nope, there are over 100 species in Minnesota and over 500 in the US.

Mayflies must have anticipated people's fear of insects. The long filaments at the end of the insect, as long as the body itself, look like something that could make a wasp sting feel good. But they are not stingers, apparently, they serve some other purpose, perhaps sensory or to distract predators. Still, one of my daughter's-in-law was almost frantic when one landed on her head.

And to compound her (misplaced) fear, when I picked up one to photograph, it waved its abdomen

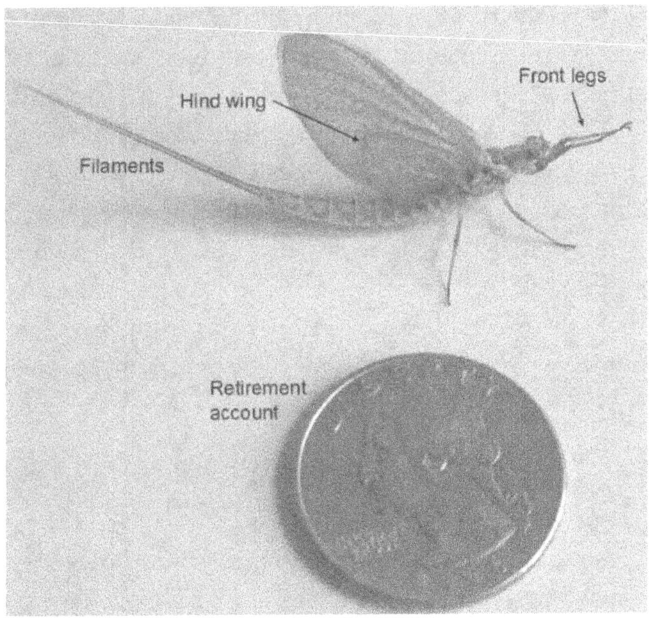

Next to last stage in the life of a giant mayfly from Leech Lake, Minnesota.

around and if I didn't know that the filaments were harmless, it appeared to be looking for way to jab me with them. The banded abdomen contributed to a fleeting thought that this looks vaguely like a wasp. Maybe a predator would assume the same?

You can convince yourself that the filaments at the end are not dangerous, but the front end looks like it could poke a hole in your finger. However, what appears to be a nasty weapon is actually their front legs, which they hold forward and together. I realized that when they're hanging on a surface, like my screens or windows, they use just their back two pairs of legs. At first, I thought, hey, how can these be insects with just four legs? I wrongly guessed those forward facing dagger-like structures were antennae, but the antennae are very small, thin and located where you'd expect them to be.

Mayflies spend 99% of their lives underwater and emerge in large numbers almost simultaneously. I'm not sure how they are so synchronized, perhaps water temperature or light conditions trigger the hatch. Remaining on my screens the next day were the shed exoskeletons (called exuvia if you wondered) from which the reproductive adults emerged. Frankly the exuvia look as though spiders sucked out their innards overnight. But no. Mayflies are the only insects that molt into a form with wings after having wings in the previous life stage. Speaking of wings, what could mayflies have in common with passenger pigeons?

The answer might be "predator swamping." Both organisms present huge numbers of themselves to potential predators, eggs and squab in the case of pigeons, and winged defenseless bags of low-fat protein in the case of mayflies. You might predict there would be a bird that was a specialist predator on mayflies. But there isn't one because there would be little for them to eat the rest of the year (unless they could also forage underwater). By having huge numbers, many females succeed in returning to water to lay their eggs because predators cannot eat them all.

Ironically, having escaped being eaten as a winged adult, the females die after laying eggs. It is also thought that a synchronous hatch of millions of individuals makes it easier for males and females to find each other, in addition to predator swamping.

Passenger pigeon colonies included hundreds of thousands of visible, easily found nests on which many different predators could feast. But the feast was short-lived, only lasting the nesting season. There was no way predators could eat all or most of the eggs during the rela-

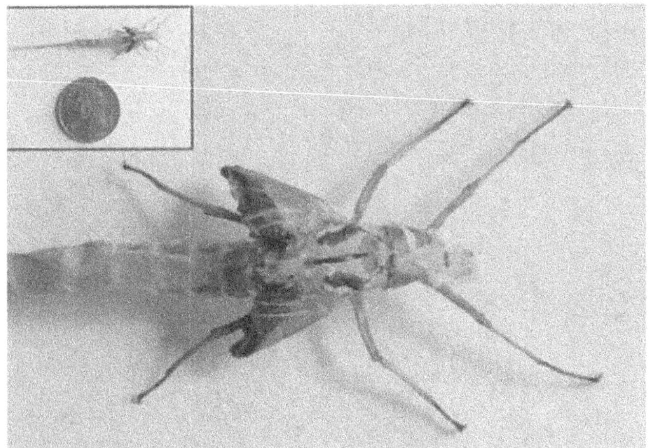

Molted exoskeleton from a giant mayfly from with final winged stage emerged; inset is full size

tively short nesting season, and like the case of mayflies, what would a passenger-pigeon-egg predator eat the rest of the year?

Hence, the term predator "swamping" refers to a strategy where vulnerability is limited in time and space, preventing the coevolution of specialist predators. Being superabundant is a sort of predator defense. Yes, some passenger pigeons succumbed to predation as do mayflies, but the level of predation is insufficient to affect overall populations.

Lastly, why "May" flies? They were probably first noticed commonly in May and the name stuck even though they are obviously not limited to that month.

77. The Worlds We Don't See: Mysophobes Beware

Cabin Talk.—Even to people who aren't nature-phobes, most can probably name a few things outside the safety of their walls: squirrel, bird, mosquito, tree, worm. But what is amazing is that inside the walls of our houses and apartments, most of us, myself included, are utterly

unaware of the equally impressive diversity of living things that we walk amongst every minute. Read below and learn what a mysophobe is, I didn't know until I learned about bathrooms and kitchens. And by reading this, you'll know why soil bacteria are found on the toilet flush handle.

I feel comfortable in northern Minnesota. I know all the birds and mammals, most of the reptiles and amphibians, the game fish but not that many other fish, quite a few insects, some big trees but relatively few other shrubs or smaller plants – land or water, some snails, and virtually no fungi or aquatic microorganisms. Actually, I am now rethinking how comfortable I should be.

Research by a team of microbiologists gives new meaning to knowing your surroundings. Gilberto E. Flores from the University of Colorado, Boulder, and his colleagues study things too small to see with the naked eye, like bacteria. In particular, they studied the distribution of bacteria on surfaces in two places we all frequent regularly, bathrooms and kitchens. The latter we hope defines cleanliness next to godliness, the former, not so much. Although we are aware of different outdoor habitats such as forests, grasslands, deserts, few have realized that different places in our homes also contain a variety of habitats, invisible to us but forming a microbial biosphere, where the microbial equivalents of predators, omnivores, and herbivores live fast paced lives. Warning: the following will be uncomfortable for any germophobes (technically speaking, mysophobia).

We know that kitchens harbor potentially harmful bacteria, but the denizens of these habitats have been relatively underappreciated owing to the difficulty in actually discovering them. In the past, one swabbed a surface and then tried to culture whatever bacteria were present. However, many bacteria do not grow in culture, and many were missed. Kind of like birdwatching without binoculars. Flores and colleagues used a new genetic technique, high throughput barcoded pyrosequencing of the 16 S rRNA gene (found in all bacteria), which reveals the actual inhabitants of the sampled surfaces. The gene sequence of each bacterium is like a barcode, which can be looked up in an international database called Genbank.

In kitchens, they found 34 bacterial groups (and likely many species), lurking unseen on 80 different kitchen surfaces, including counters, cabinet handles, microwave touch screens, fans, stoves and refrigerators. For comparison, in the Minnesota winter I have fewer than 10 species of birds at my feeders. In general, kitchen surfaces with the most bacteria were on the outside of stove exhaust fans and the floor (so much for the "three second rule"). Refrigerator door seals were also frequently contaminated. Sinks didn't have all that many bacteria, but because they are usually moist, they did harbor bacterial diversity.

The bacteria are known to occur in food and on the human body and are known to be able to survive on exposed surfaces for long periods of time, up to two weeks. The good news is that most of the bacterial types known to cause human illnesses were relatively rare. The bad news is that they were widely distributed in the kitchens, even on surfaces unlikely to come into direct contact with raw food. For example, "Campylobacter," which causes intestinal problems, was found on cabinet handles or the microwave touch screen panel. Kitchen users probably handled raw chicken and then touched these surfaces. I'm not a committed germophobe but the first thing I did was to look closely at my microwave touch screen. Sure enough, I couldn't see anything, but I attacked it anyway with a damp cloth and soap.

Of course, one would not want to use a very powerful anti-bacterial soap because it would challenge the bacteria, and natural selection would lead to new, more powerful mutant strains on my microwave. Then, I'd have to get a more powerful soap, which would work only for a short time owing to the evolution of new and more resistant bacteria. Darwin would have loved it – an evolutionary battleground, unseen, in your own kitchen.

Interestingly, different places within a single kitchen were more similar than the same areas (e.g., sinks) of different kitchens, probably because of unique cooking and cleaning habits of the occupants.

Given all this good news from kitchens, I didn't even want to hear about bathrooms. That turns out to be pretty much true, but I will say that it will give you greater appreciation for your immune system. The researchers this time swabbed 10 surfaces in 12 public restrooms (6 men's, 6 women's) and identified bacteria through the same genetic testing that they used in kitchens. Surfaces associated with toilets (seat, flush handle), restroom floors, and surfaces routinely touched by hands (door in/out, stall in/out, faucet handles, soap dispenser) were the main areas that tested positive for bacteria.

A total of 19 major bacterial groups containing several hundred different bacterial species were found in the restrooms, and those associated with human skin were found on all surfaces. You won't be surprised that the most frequent bacteria on toilet surfaces were gut bacteria. Contamination occurs via direct contact or indirectly when the toilet is flushed and water splashes about. Floors harbored the most diverse bacterial communities, many of the species being traced to soils, likely brought in on the soles of shoes.

The most amusing finding of the study, if amusing is even an apt characterization, is that a large number of soil microbes occur on toilet flush handles. This seems odd until you realize that it is common practice among high end germophobes to operate the flush handle on the toilet with the bottom of their shoes instead of touching it with their hands. Unfortunately, I'm getting to the age where I'd have to balance (literally) the possibility of avoiding touching the handle with a lower back injury.

Was there a gender bias in the distribution and abundance of bacteria in men's and women's bathrooms? I admit to a preconceived bias, having observed teenage boys grow up. But there were no major differences, which was attributed to the fact that other studies that show that college students, the most frequent users of the 12 restrooms they studied, are not particularly diligent about hand washing. However, perhaps lending some validation to their methods, bacteria specific to women's reproductive tracts and found in urine were more common on toilet seats in women's restrooms. They summarized: "the prevalence of gut and skin associated bacteria throughout the restrooms we surveyed is concerning [because bacteria] could readily be transmitted between individuals by the touching of restroom surfaces." Seems to me like one of the biggest understatements since Noah remarked (purportedly) "It looks like rain."

I might know where every knife, serving utensil, and sauté pan resides in my kitchen, but the concept of them embedded amongst a diverse microbial ecosystem is foreign to me. It never occurred to me that my kitchen might be as ecologically diverse as the outdoor ecosystems I know (or think I know). As the authors noted, "More than ever, individuals across the globe spend a large portion of their lives indoors, yet relatively little is known about the microbial diversity of indoor environment." As I pondered their results, it occurred to me that it's a marvel we aren't sick all the time. The next time you need to toast something, toast your immune system, but maybe first wash your hands (or in the case of my late father, a happily committed germophobe, for the 90th time that day). And if you're in the mood for seeing biodiversity, but don't want to put your shoes on, grab some swabs and a microscope and take a nature walk through your house.

78. How Do You Say 1,000,000,000,000,000,000,000,000,000,000?

Cabin Talk.—Other than "No idea, but if it's money I'll take it" you probably did what I did and tried to remember from your school days the progression: million, billion, trillion, ah crap, what's next? Before I give away the answer, perhaps "Why did this come up" would be relevant. Well, the answer is not the subject of most articles about the outdoors, but it has much meaning to people who love being outside.

In general, I think we know a lot about how much nature there is around us. We have pretty extensive catalogs for most of the earth's areas for birds, mammals, amphibians, reptiles, and most vascular plants. However, we know very little about things too small to see. What if I told you that with the right optics, you could see 140 different species of deer in Minnesota, not just one?

In the news there has been a huge focus on things too small to see with the naked eye. We now know that if you look at your average deer, for example, it has more cells belonging to micro-organisms than it does those belonging to deer. In the case of humans, for many years microbiologists repeated a figure of 10:1 microbial cells to human cells. That doesn't necessarily mean that if you removed all the human cells you'd still see our shapes. Microbial cells are much smaller that human cells, and in turns of volume, our microbial cells would fill a half gallon jug. Recent writers have noted that the 10:1 ratio was almost myth-ish, and cell biologists now think the ratio is closer to 3:1. Still, the human body is not "all human" nor is the body of any other animal mostly made up of cells of that animal.

Even more recent has been the realization that each of us has different gut "microbiomes". The microorganism composition of our gut can influence things from how healthy we are to our weight. A developing medical treatment for *Clostridium difficile*, a bacterium that causes diarrhea to life threatening inflammation of the colon, is FMT. If you're eating you should skip the rest of this paragraph, because FMT stands for fecal microbiota transplants. Yes, fecal extracts from one person are injected into another to "fix" their out-of-whack intestinal microbiomes. Given how little we know about gut biomes and health, maybe a lot of wildlife diseases have a root in unnatural gut compositions?

Deer enthusiasts will now understand that the shift in food from summer to winter is accompanied by a shift in the gut microbiome. The reason for recommendations against feeding deer hay in the winter is because the gut bacteria deer have in winter are those that digest browsed food items like twigs, and in some cases, eating too much hay in winter with the wrong gut critters can kill a deer.

Ok, back to our big number. It comes from ongoing work that is cataloguing the viruses of the world's oceans. Prior to the new work, a total of 39 different ocean dwelling viruses were known to scientists. After this work, the number grew to 5,476. In just 20 gallons of ocean water, up to 3 trillion viruses exist. Think about that. Could we miss seeing 3 trillion squirrels?

Why had so many virus species been missed? Turns out that one virus pretty much looks like another, so seen one, seem em all. But in the 1990s, scientists realized you could add some chemicals to a pot of viruses, break them down, and recover a DNA virus soup. The DNA of the virus is like a bar code and it turned out there are lots of really different ones that look alike and were thought to be the same species. 99% of the viruses discovered by DNA comparisons are new to science.

What do these viruses do? They attack bacteria and other marine life, just like the influenza (flu) virus attacks us. But their role in the ocean ecosystem is just beginning to be unraveled.

I thought this was interesting for a couple of reasons. First, we tend to underestimate the importance of things we can't see. Turns out that the oceans have been harboring 140 times more viruses than we thought. The scientists estimated that there are 1,000,000,000,000,000,000,000,000,000,000 virus particles in all the world's oceans. So, what's in a number?

The reason viruses can be so deadly is because of their huge numbers. For example, if there were 100 trillion virus particles, and your drug killed 99.999% of them, there would still be 100 million that survived because they have the right mutations, which of course is why it is so difficult to combat viruses, and you get a new flu shot every year.

The coming years should provide some very new insights about the biology of all creatures when we better understand the microorganisms with which they co-exist and co-evolve. I wouldn't be surprised if there are some totally unexpected findings.

And finally, the word of the day is nonillion.

79. Prince Was A Hunter And Liked Seafood

Cabin Talk.—I thought I'd end with a look back in time at a youth named by his discoverers as Prince. No one knows his actual name. I found the story surrounding his life and demise to be a fascinating glimpse of what we'd be doing, were it not for the intervening 24,000 years since he was alive.

Being from Minnesota, the name Prince immediately reminds me of a performer. However, this note is

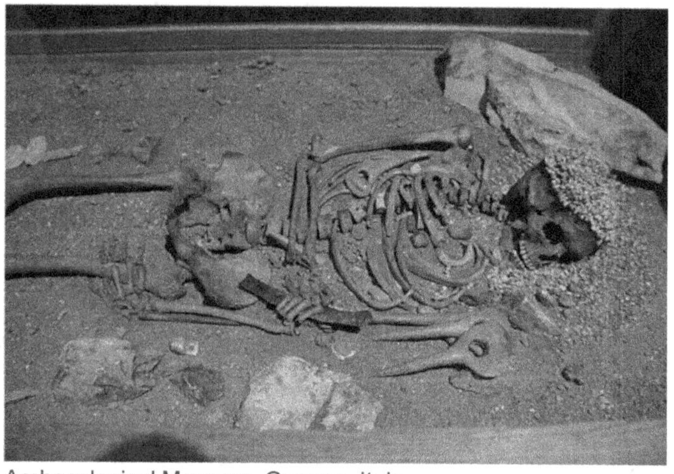

Archaeological Museum, Genova, Italy

not about that Prince, rather a young man named Prince

that died 24,000 years ago (see image) after what was thought to have been an unfortunate encounter with a bear or cave lion. At the time of his demise, he was between 15 and 17 yrs old and stood about 5'6" tall. His remains were found in 1942 about 22 feet below the surface of the entrance of the Arene Candide Cave, which is located in extreme NW Italy. The skeleton is complete and very well preserved, which imo, means it is unlikely that a bear or lion was the culprit in Prince's death, because I would think that either predator would have consumed his body.

What's Prince's significance to hunting and seafood? People who have studied the skeleton concluded that he had very strong arms, especially his right arm, likely for throwing spears, and well developed leg bones that were probably a result of "continual and prolonged exertion" when hunting on foot. Future archaeologists will have no trouble identifying my hunting style, a ground blind.

At least in death, Prince had a hat or headdress of small, perforated shells (I'm guessing snails) and deer incisors (the hat was reconstructed atop his skull after the discovery of the remains). Looking good was apparently a priority, although teenage boys can take fashion to the next level, maybe his parents were rolling their eyes. In his right hand was a 9-inch blade made of flint (microscopic crystals of quartz), a hard substance used for thousands of years to make tools, and later to start fires (it sparks when struck against steel – think flint-lock rifles).

Archaeologists think the flint came from the Forcalquier basin in southeastern France, about 80 miles from the cave. Although I assumed that this blade was for killing or processing game, or self-defense, archaeologists suggest it was a gift for the grave because it showed no signs of use. People of the time likely used similar blades for hunting, but this was made specially for Prince to take into the great beyond (although it stayed behind).

Many other interesting items surrounded his final resting place, including a bracelet made of shellfish, pendants from mammoth ivory, perforated batons made from elk antler carved with decorative lines and notches, and two mammoth ivory buttons next to his knees. A necklace consisting of small shells ends with cypreae shells, a reasonably large sea snail or cowrie: a shell frequently linked to female fertility. His grave was covered with ochre, which gives the skeleton a reddish coloration.

I was fascinated by the perforated antler pieces or batons, and like anyone, can come up with some potential uses, ranging from pure decoration to some kind of stabbing weapon. Others suggest they were used to straighten arrow points and spears. Apparently, there are few objects like this in European archaeological sites, and the batons have also been considered to be royal sceptres, or a wand held in hand by a high-ranking person (e.g., a Prince, but not the one from Minnesota). It appears that Prince commanded respect, certainly something I didn't have when I was 16. Prince was either royalty or a first class hunter.

Given the grave artifacts, there seems little doubt that people of Prince's day hunted deer, mastodon and other creatures, and likely scavenged some too (I kind of doubt that a 9 inch blade would bring down a mammoth). Chemical analysis of Prince's bones indicate that he ate wild game, and analysis of isotopes from his teeth revealed that approximately 25% of his diet consisted of marine protein, which makes sense given that the cave is close to the sea coast (remember, you are what you eat, at least to some extent).

Staring at his image brings to mind many questions. What was it like to hunt for food back then? How did he or relatives get the deer/elk antlers? Did they hunt woolly rhinos and European bison? How did they hunt, how common were game animals, did they worry about the wind, were they camouflaged? I am guessing that kills were made from pretty short distances because the atlatl, which allows the hunter to sling a spear with great force, didn't appear until around 21,000 years ago, after Prince's unfortunate demise. What was the strategy when a mammoth lumbered within range, was it every man for himself or was there a meeting?

When we hunt, we're not being hunted at the same time (mostly). As Prince stalked prey, did he watch out of the corner of his eye for cave lions, wolves, cave hyenas, or bears? What would you do if you were just about to jump a deer and a cave lion attacked you, and you could only get off one arrow; I'd shoot at the predator. My guess is you never hunted alone, especially, if like me, you wear glasses. And then recall the common adage, don't be the slowest one in your group.

I can't explain why I was so drawn to this image and why I wanted to learn what I could about his life. I suppose it's because I see a little of him in all hunters including myself. We hunted to survive then, and some of us carry on the hunting and fishing tradition, 24,000 years later. Long live the tradition.

80. Dogs And Traps: A Cautionary Story

Cabin Talk.—Dog people usually think of their dogs as part of the family. We keep pictures of dogs from the past and we talk about what this dog would have done in this situation. We tell "remember when" stories. I look at old pictures and hope he's on point somewhere. I take a long time to get over a dog we've had to put down. A friend recently had a dog killed in front of her by a passing car; I can't imagine the horror. So, we do everything we can to protect our canine companions, but sometimes unanticipated events get in the way.

A friend was in a Minnesota state forest legally cutting down a Christmas tree. His two drahthaars were nearby. One began making unusual noises, and my friend assumed that the dog had encountered a racoon or skunk, which is usually bad for them. He brought his other dog back to his truck and went to investigate, only to find his 6 yr old drahthaar caught in a 220 Conibear body trap around its neck and head. He figures the dog had been in the trap for almost 10 minutes by the time he got to it, but fortunately, the dog was not caught in a way that would have crushed his trachea and ended its life. My friend struggled with the trap and finally got it released. Although he carries zip ties in his hunting vest to release a dog caught in a trap, he was not hunting and his vest was in his truck, several minutes away—time he didn't have to manipulate the trap to prevent the dog from dying.

After struggling for what seemed like forever, he was able to reduce the trap's pressure on the dog's neck and eventually remove it entirely from the dog. The trap had no identifying information on it and appeared to have been broken at a snap at the end of a cable that presumably had been attached to a restraint. Possibly, as required, it was in a box labeled with the trappers information and inset at least seven inches, but even if so, the trap obviously wasn't far enough removed to prevent the dog from near disaster. The dog seems to have luckily avoided permanent damage although there was swelling around his eye. A dog of lesser physical stature than a Drahthaar would likely not have been so lucky (my old English Setter for instance).

Online sources have suggestions for hunters whose dogs encounter body-grip traps. For example, Minnesota DNR says "If a dog is caught in a body-grip trap you have a brief time to safely release the animal. It is important to act as quickly as possible." Although well intentioned it does not help that the DNR encourages dog owners to practice releasing techniques—removing a trap from a dog in the heat of a moment is different from setting and unsetting a trap on your garage floor. And they mention that to open the trap, use your hands or a "setting device." Also, they demonstrate methods for using ropes, leashes or zip ties to release a dog in a death trap—probably a good thing to know, but the onus is on dog owners to deal with the consequences of their dog caught in a lethal trap.

Dog owners should not have to do this. I should not have to carry a "setting device." Nothing I do in the outdoors will endanger the life of your dog. I should not have to be prepared to release my dog from a potentially fatal experience in a trap, never. Even one dog killed by a trap is too many. There needs to be better safeguards.

The MN trappers association writes to their members: "Please be very careful of where you set traps, especially baited body grip traps. Stay away from trails and places where people might be walking dogs." Further advice: "we can certainly stay away from public places where grouse hunters and trail walkers frequent." Anyone who has hunted even once with pointing dogs knows they do not stay on trails, their job is to hunt a wide berth and find birds. A trap would have to be at least 100 yards off a trail to be out of range of my dogs.

I have always been supportive of trappers, as has my friend who had this experience. And there have been changes in trapping regulations, like delaying the season so that trapping occurs when fewer grouse hunters are out and requiring traps to be elevated (which trappers say makes them ineffective) or in boxes designed to keep dogs out. A 220 conibear in a box baited with meat is something that almost any hunting dog would investigate, and in this case, be able to get their head far enough into the box to have the trap grab their neck and kill them. My drahthaar pinned himself under a bed trying to retrieve a ball, he'd go anywhere he thought he had a chance for meat.

The fact that dogs are still caught in traps means the safeguards are not always working. Yes, state forests are public areas open to recreational opportunities for everyone. But, if your idea of recreation can kill my dog, there's a problem.

81. Epilogue

As a cabin owner in northern Minnesota, I know that in just about any week there will be a day where it's too cold, too rainy, too windy, a blizzard, or too hot (less likely) to be outside. I put these essays together thinking of those days. I have learned a lot in my time as a professor, but I'm continually amazed and how much I continue to learn, and how much I didn't know I didn't know. I hope these stories and ideas give you something to do on those inclement days at the cabin that you find rewarding.

Acknowledgments. D. E. Andersen taught me about hunting dogs. R. Drieslen, J. Janovy, and T. Spielman guided my writing. My wife Susan tolerates, even joins, my addiction to things outdoors. My sons have added an inestimable joy to our adventures. Most of these essays appeared in Outdoor News, and I'm grateful for their support. Many images came from Pixabay.